civics+ citizenship
geography
economics+ business

contents

Environmental change and management

Geographies of human wellbeing

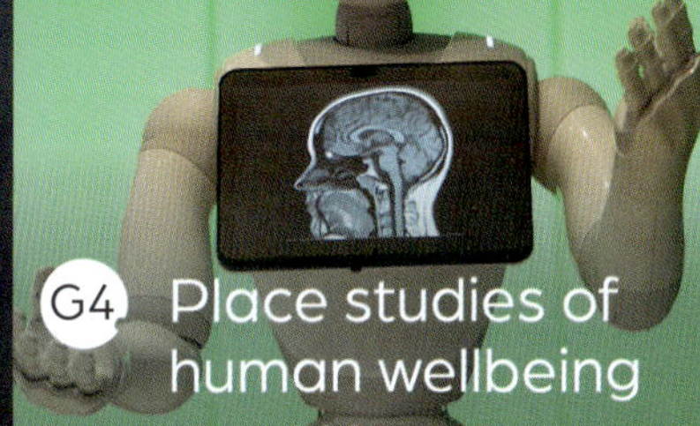

Geographical concepts + skills

Good Geography 10

Victorian Curriculum

1st edition

Danielle O'Leary

Phillip O'Brien

First published in 2022 by Matilda Education Australia,
an imprint of Meanwhile Education Pty Ltd

Melbourne, Australia
T: 1300 277 235
E: customersupport@matildaed.com.au
www.matildaeducation.com.au

Publisher: Catherine Charles-Brown
Publishing director: Olive McRae
Copy editor: Kate McGregor
Project editor: Kate McGregor
Proofreader: Kelly Robinson
Indexer: Max McMaster
Text and cover designer: Jo Groud
Typesetter: Paul Ryan
Illustrator: QBS Learning
Cartography: Stewart Adrain at Custom Mapping Services
Production controller: Sarah Blake
Permissions researchers: Samantha Russell-Tulip, Hannah Tatton
Cover images photographed by:
Justin McManus/*The Age*. Reproduced with permission from Fairfax Media (Nine Publishing).

National Library of Australia Cataloguing-in-Publication data

Author: Sally Elliott, Aleryk Fricker, Damien Green, Ben Lawless, Phillip O'Brien, Danielle O'Leary, Natalie Shephard, Ilja van Weringh
Title: Good Geography 10 Victorian Curriculum
ISBN: 9780655091318

A catalogue record for this book is available from the National Library of Australia at www.nla.gov.au

Warning: It is recommended that First Nations peoples exercise caution when viewing this publication as it may contain names or images of deceased persons.

Printed in Malaysia by Vivar Printing

Dec - 2023

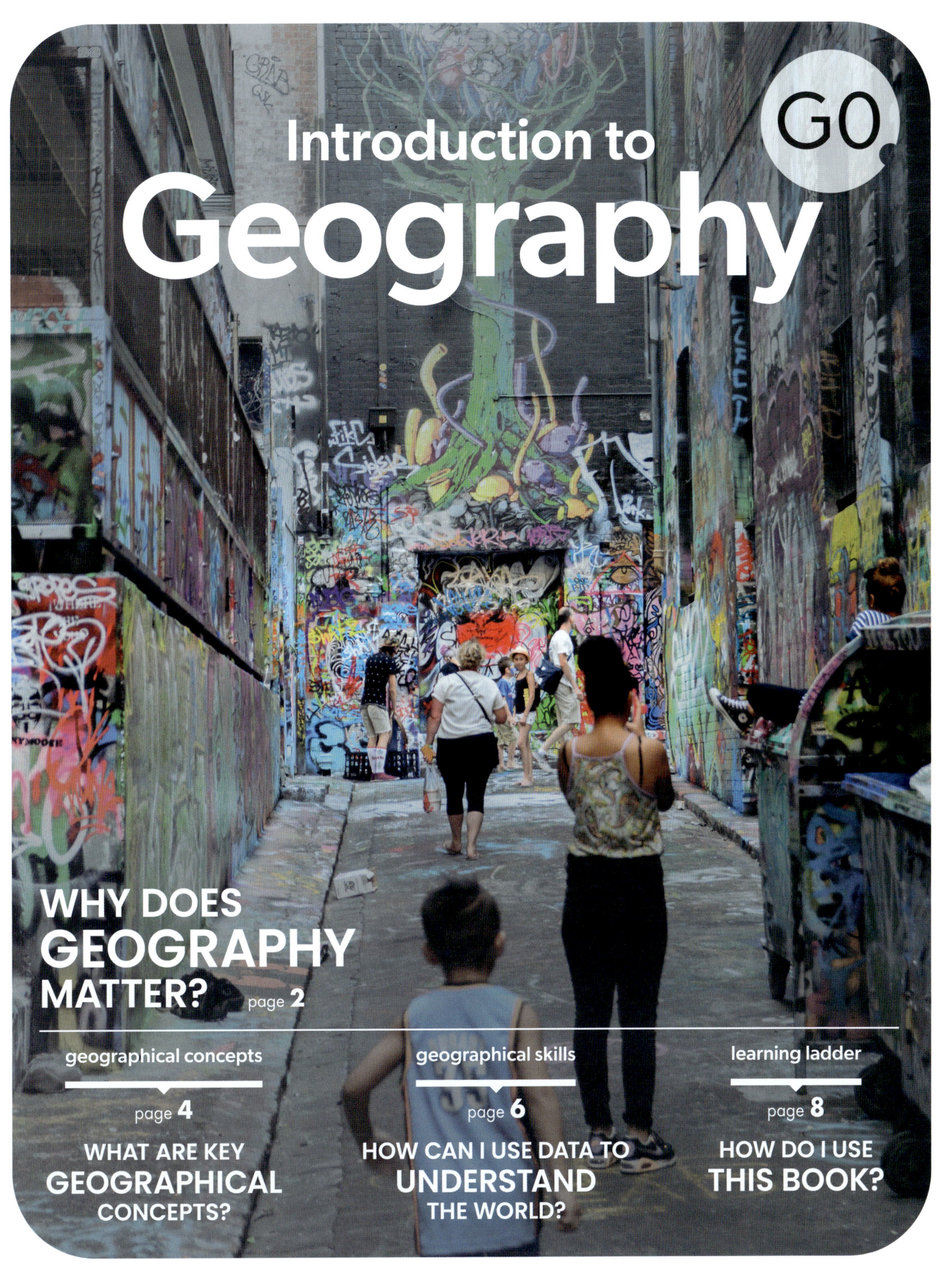

G0 Introduction to Geography

Why does Geography matter?

Geography is the study of the world: its characteristics, its patterns and how it changes over time. In Geography, we focus on two main factors: human activity and natural processes. We are interested in how humans influence the *space* in which they live and, in turn, how the natural environment influences how humans live.

Previously, you learned how to interpret sources such as maps, images and graphs, how to explore spatial patterns and how to conduct fieldwork to answer research questions and find interconnections between phenomena. As you move towards VCE, you will now build on these skills and apply your knowledge from earlier years of study to understand new case studies, make predictions, analyse processes and further your connection with communities on various scales. You are the future of our world and Geography is your key to understanding it!

How will Geography help me in the future?

Have you ever wondered when you will use the skills you learn in Geography after finishing school? Beyond developing a key understanding of the world around you and being able to engage in society as a knowledgeable, informed and empathetic global citizen, you will be able to use maps, apps and **geospatial technologies** more effectively in everyday life. You will also be able to apply theoretical concepts to new situations and adapt your ideas to predict future outcomes. You will also be able to think critically, reflect and analyse processes, and solve problems. You may not choose a career that is directly related to Geography, but the skills you are learning today are applicable to many things you will do in the future.

Source 1

Possible career avenues directly linked to geographical studies

Source 2

When Swedish school student Greta Thunberg staged a protest in August 2018 holding a sign that read '*Skolstrejk för klimatet*' (School strike for climate) she motivated tens of thousands of students globally, like the student in this photo, to protest for **climate** action.

Learning ladder G0.1

1 This year, you will explore environmental change and management, and human wellbeing. List some of the topics you explored in Years 7–9 that may help you in understanding new case studies and processes.

2 What is meant by the word 'phenomena'? What is an example of a geographical phenomenon?

3 Using your knowledge of current events, brainstorm some relevant case studies that connect to the topics of environmental change and management, and human wellbeing.

4 Consider what you would like to be when you finish school. List ways that geographical skills may help you achieve in your chosen career.

What are Key Geographical Concepts?

Previously you used SPICESS to write geographically and explore phenomena. In VCE, the 12 **Key Geographical Concepts (KGCs)** replace or extend these terms.

The KGCs allow you to continue to improve your geographical writing skills and practise thinking more critically about world processes. The KGCs appear in *italics* in the Learning ladder to prompt you to consider them in your answers.

1 Place

The concept of *place* allows humans to identify the location or position of something within a space. We can identify place through describing the relative or absolute location of that area. This is Luwu Timur, South Sulawesi in Indonesia, in the southern hemisphere.

2 Scale

Scale usually refers to the size of something. Scale can be literal, such as the scale on a map using quantitative data to represent the size of a real-life object. In this photo, large-scale vegetation clearing has taken place for palm oil farming.

3 Distance

Distance refers to the amount of space or time between two places. Temporal distance refers to the time it takes to get from point A to point B. For example, a farmer may take 30 minutes to travel to their fields from their village. Spatial distance refers to the physical distance between points A and B. For example, it may be 2 metres between each palm oil tree.

4 Distribution

Distribution refers to the way phenomena are spread across a space. For example, the palm oil trees could be described as being in a uniform or linear distribution, while vegetation in the forests may have a clustered or random distribution.

5 Movement

Movement describes a change in location due to a particular process. For example, the natural vegetation in this forest may be logged and transported to other regions for pulp or timber.

6 Region

Region refers to an area or group of countries that have similarities. A region may be based on climate, such as a tropical rainforest or desert region, or based on location, such as Southeast Asia.

7 Change

Change refers to how a *place* is altered by shifts in the environment or to meet the needs of humans. Change can be positive or negative and can occur over short or long periods. This place has seen significant change over time, mostly due to human interaction and the removal of natural land cover.

8 Process

A *process* is a series of events or steps. Processes can be both natural and human, and form a range of interconnections that impact people and place. For example, we could explore the processes that have led to the dramatic changes in this place.

9 Spatial association

Spatial association describes how two phenomena occur in the same place, at the same time, for a particular reason. It is a way of describing how one phenomenon is interconnected with another. We can describe spatial associations using terms such as 'strong' or 'weak'. For example, there is a strong spatial association between mountainous topography and low population density.

10 Sustainability

Sustainability is the concept of maintaining and preserving resources, environments, societies and economies for future generations. Sustainability means balancing environmental and economic needs. It is not as simple as asking palm oil farmers to stop, as communities rely on these crops for their income. Our job as geographers is to collaborate with other specialists such as economists and scientists to create sustainable solutions for society and the natural world.

11 Environment

When we think about the *environment*, we often picture a natural space. However, with increasing populations and urbanisation, it is also crucial to consider human environments. An environment is made up of all the living and non-living factors within a space.

12 Interconnection

Interconnection is the idea that two things or phenomena are related, interact or are linked in some way. For example, there is a strong interconnection between the success of a region's harvest and its economy.

Learning ladder G0.2

1. List two similarities and two differences between the SPICESS and KGC terms.
2. Explain how your local *place* has *changed* over time.
3. Describe the *spatial association* between human population increase and agricultural output.
4. '*Sustainability* is *interconnected* with more than just ensuring the longevity of the environment.' Discuss as a class using examples (hint: think SHEEPT).

G0.3

How can I use data to understand the world?

As Geography spans both Humanities and Science, geographers use both **qualitative** and **quantitative data** to describe processes and changes in places over time. These different data forms are important for providing context, quantifying changes and illustrating processes.

As you move into higher levels of geographical studies, it is important that you provide data in your responses, reports or projects. Data could be in the form of descriptive examples, case studies, photographs, numerical statistics or infographics. Data should be integrated into your writing.

While writing a list of 'fun facts' may have been suitable for previous years, you now need to practise using data as part of your response as you would for a quote in English or an equation in Mathematics. The PQE and SHEEPT guides continue to be good structures to help you use evidence in your responses.

Maps

Maps can provide you with a range of data including distribution, movement and scale. You could gather quantitative data from maps by using the scale or legend, or describe processes and phenomena through specific examples. Different maps can be used to tell different 'geographical stories'. For example, in Source 1, many coastal cities are expected to be inundated if the sea level were to rise by more than 3 feet (91 centimetres). Low-lying areas in San Bruno are particularly vulnerable to flooding. This data provides key information to policy makers about how to respond to future climate events.

Other maps such as cartograms can sometimes be confusing because they are distorted by data, but they can be very informative when used correctly. In Source 2, both the choropleth map and the cartogram use the same legend, but display different information. For example, while the choropleth map colouring may suggest a higher number of votes went to presidential candidate Donald Trump, when studying the cartogram, we can see that the states with the highest populations (which are larger than the others) supported Joe Biden, meaning he won the majority of the votes (51.1 per cent compared to 47.2 per cent) in the 2020 US presidential election.

Source 1

Example of a GIS output highlighting regions at risk of coastal flooding because of climate change-induced sea level rise

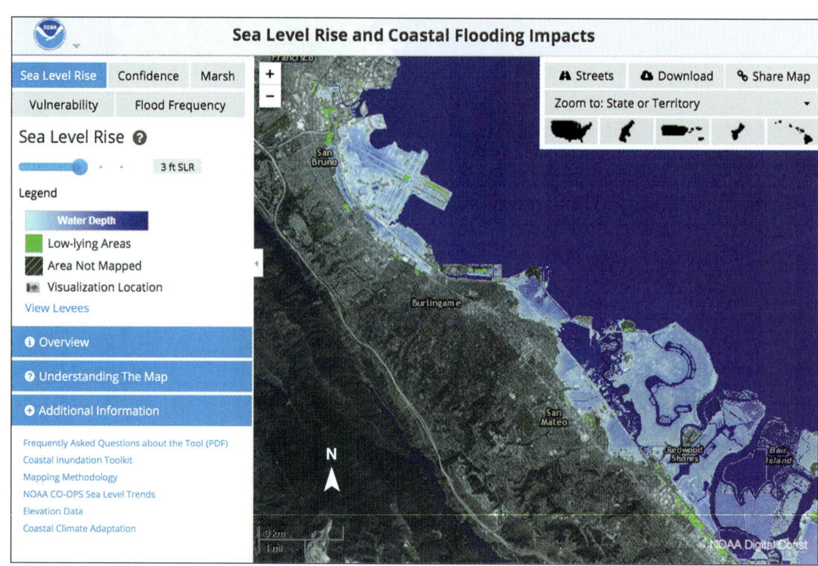

Source: climate.gov

US Presidential Election results 2020

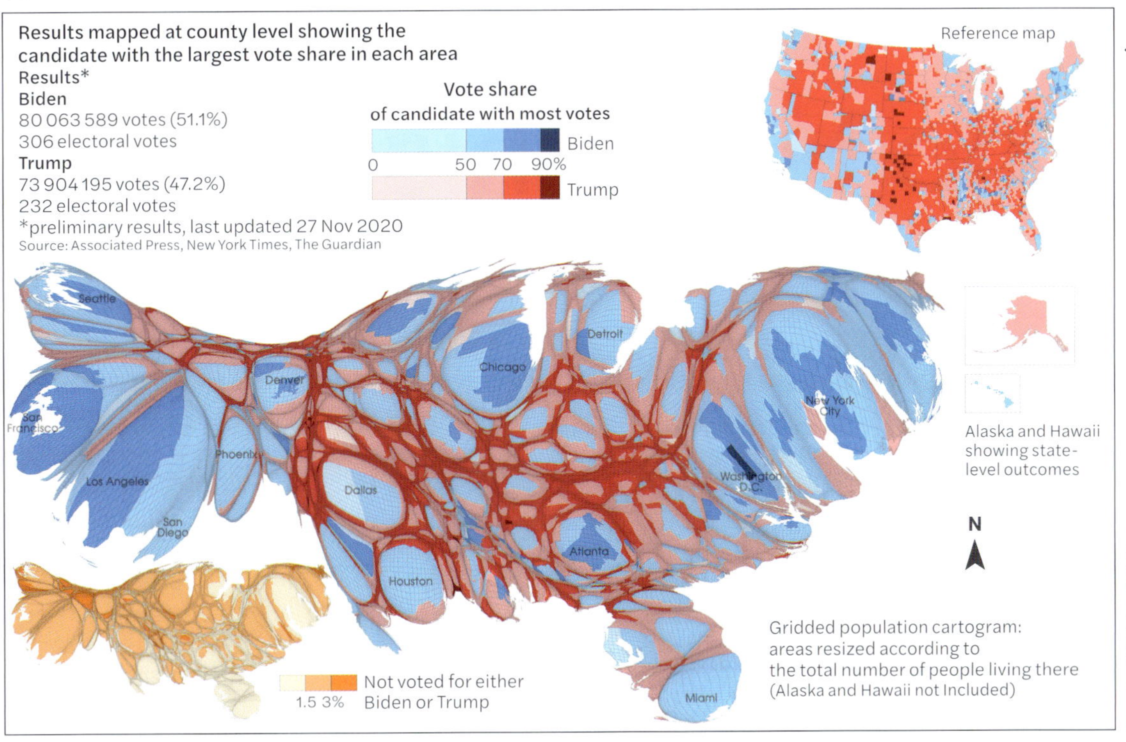

Source 2

Choropleth map and cartogram showing the same information but telling different stories

Graphs

Graphs provide quantitative data that can help identify patterns and changes over time. The axes and data points allow you to compare and contrast. For example, in Source 3 we can see that globally the unhappiest female age group is 30–55 years old, with a happiness score of just over 5 out of 10. Graphs may also help you qualitatively explain data or draw conclusions. For example, women in the 30–55 age bracket may rank as least happy because they are busy with work and family, whereas women of other age brackets may have more time to engage in recreational activities.

Average female self-reported happiness by age, 2016–18

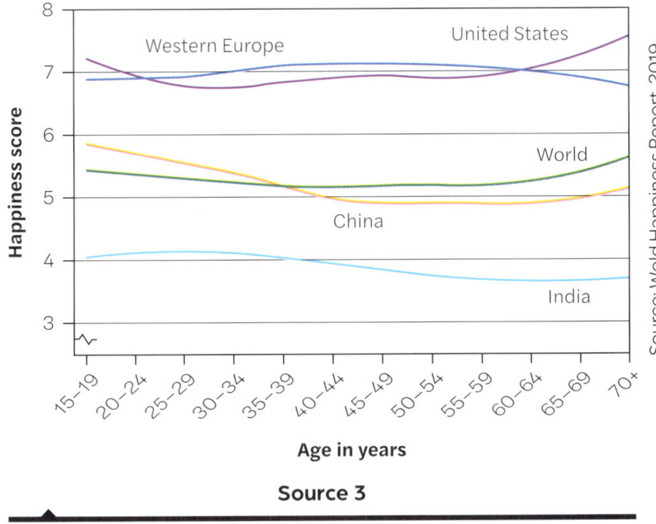

Source 3

This line graph shows changes in happiness over time for women.

Diagrams or sketches

Diagrams and sketches display qualitative and quantitative data on processes or changes. Annotations on sketches can help to describe different steps in a process or allow the reader to understand the symbols used. The scales or legends completed as part of a sketch, especially those done during fieldwork, provide estimated quantitative values as evidence of change in a particular place at a particular time.

Learning ladder G0.3

1. Why is data important in Geography?
2. Source 2: Suggest how map type can be important to convey geographical data, with reference to the US presidential election.
3. With reference to examples from sources in G0, outline the differences between quantitative and qualitative data.
4. Would quantitative or quantitative data be more suitable as evidence for the following scenarios?
   a. Investigating the climate of a particular *place*
   b. Researching how people describe their level of happiness at a given time
   c. Highlighting *change* over time in an urban *environment*

How do I use this book?

Good Geography will help you thrive as you move through the Level 10 Geography curriculum and to demonstrate your progress in every lesson. This book explores two geographical topics: environmental change and management, and human wellbeing. You will also find a Fieldwork section and a Geo How-To skills section. The Geo How-To section is vital and you should refer to it often.

Climb the Geography Learning ladder

Geography is a skills-based subject. This means that learning content alone is not enough to give you a geographical understanding. In order to be a geographer, you need to be able to understand question terms, write geographically, read a variety of data, interpret data and conduct and formally report your own research.

Each chapter in this book begins with a Learning ladder. This is one way you can take control of your own learning of geographical skills. It lists the five geographical skills you will be learning and five levels of progression for each of those skills. Read it from the bottom to the top. As you progress through the chapter, you will climb *up* the Learning ladder.

Each skill described in the Learning ladder is a higher progression than the one below it. To be able to accomplish the higher-level skills, you need to be able to master the lower ones. Practising activities at all the levels will help you develop senior geography skills, such as evaluating. This approach is called developmental learning and puts you in charge of your own learning progression.

Source 1

The Learning ladder helps you to take charge of your own learning.

learning ladder

Step	Spatial distributions and patterns	Patterns and interconnections	Changes and implications	Communicate data	Digital and geospatial technologies
step 5	I can identify multiple spatial distributions and patterns	I can interpret causes of patterns and interconnections	I can interpret data to quantify predictions based on research	I can evaluate the success of research methods	I can draw conclusions from geographical information in digital and geospatial technologies
step 4	I can use data to support exceptions to spatial distributions and patterns	I can use relevant sources to research further reasons for patterns and interconnections	I can make predictions and outline consequences of change over time	I can organise data collected according to relevance for a research question	I can manipulate data using digital and geospatial technologies
step 3	I can describe spatial distributions and patterns	I can use data to support explanations of patterns and interconnections	I can explain the causes behind the change over time in a place	I can filter collected data	I can access and use geospatial technology platforms such as GIS
step 2	I can use data to quantify spatial distributions and patterns	I can explain patterns and interconnections	I can describe how places have changed over time	I can successfully use data collection methods	I can construct paper maps using correct cartographic conventions
step 1	I can identify spatial distributions and patterns	I can provide short explanations for patterns and interconnections	I can identify that changes occur in the characteristics of places over time	I can list primary and secondary methods useful for my study	I can interpret different map types using cartographic conventions

Check your progress

Each chapter is divided into sections. Each section is designed to cover one lesson, but your teacher might decide to spend more or less time on a particular section. At the end of every section, you will find a block of questions to help you check your progress.

1 Show what you know

These questions ask you to look at the content you have learned and to show your understanding of it by listing, describing and explaining.

2 Learning ladder

These activities are linked to the Learning ladder. You can complete one of the questions or several of them, depending on your progress. Throughout a chapter you will complete an activity for each ladder level.

Place studies

Geography is the study of people and places. Geographers use examples of processes or events that have occurred in the past to predict what might happen to other places in the future. Chapters G1 and G3 are centred around the background information you will need to understand in order to apply your skills to 'place studies' in Chapters G2 and G4. When you notice this symbol on a spread, it means that there is a place study available that is connected to what you are learning about. You may wish to turn straight to that page to learn more or to begin to apply your understanding.

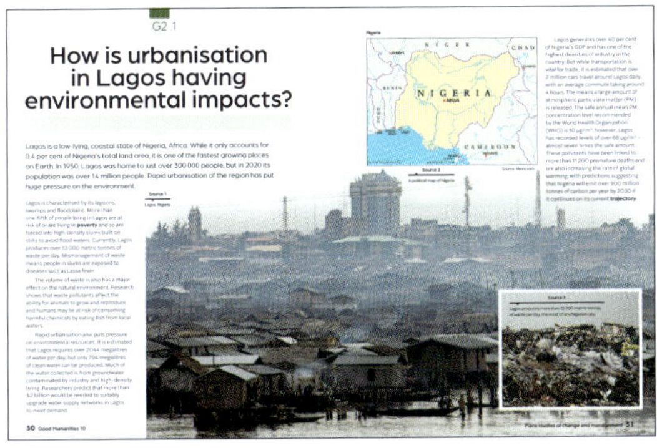
G2.1

How is urbanisation in Lagos having environmental impacts?

Source 2

Place studies are an important part of your Geography course.

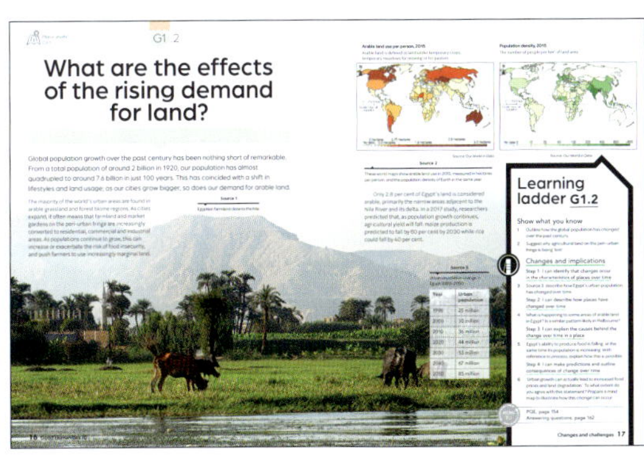
G1.2

What are the effects of the rising demand for land?

Learning ladder G1.2

Source 3

Check your progress regularly. You can attempt one or all of the Learning ladder questions. Challenge yourself!

civics+ citizenship **economics+ business**

The study of Geography can be enhanced by the study of Civics + Citizenship and Economics + Business. In Chapters 1 and 3, you can study a Civics + Citizenship and an Economics + Business lesson. School is busy and you have a lot to cover, so placing important Civics + Citizenship and Economics + Business content next to relevant Geography lessons will help you connect your learning.

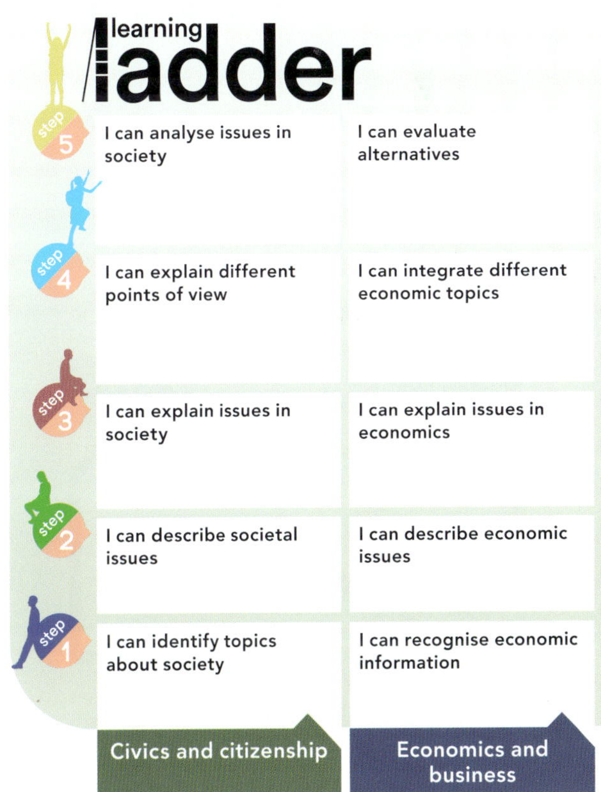

Geo How-To

At the end of the book, you will find a skills chapter called Geo How-To. This chapter explains how to perform each skill and how to answer geographical questions. Refer to it often, especially when answering the Learning ladder questions and Masterclass activities.

In addition, the Fieldwork chapter explains the skills you need for hands-on research and gives you several suggested tasks.

Source 4

Refer to your Geo How-To chapter to find out how to perform each skill.

Masterclass

At the end of each chapter is a review section, with questions organised by the steps on the Learning ladder. You can complete all or some of the questions, depending on your progress. By the end of every chapter, you will have had the opportunity to:

- learn new content knowledge
- practise how to list, describe and explain
- practise how to analyse sources, solve problems and create maps or graphs
- have two or more opportunities to practise every step of the Learning ladder.

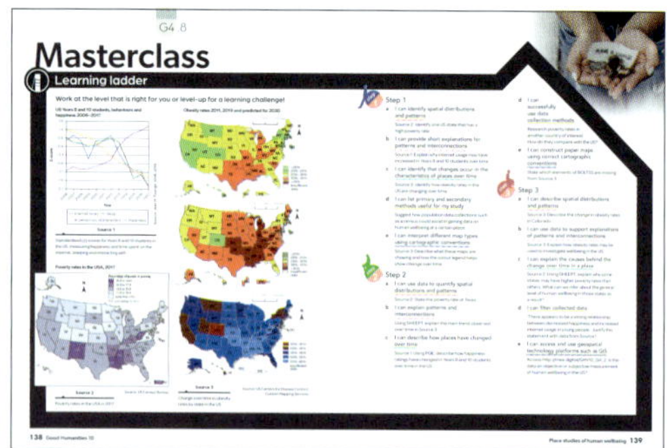

Source 5

The Masterclass is your opportunity to show your progress. Take charge of your own learning and see if you can extend yourself.

Capstone

After you complete a chapter, it is time to put your new knowledge and understanding together for the capstone project to show what you know and what you think by applying your new knowledge to the capstone project. In the world of building, a capstone is an element that tops off a building or wall. That is what the capstone project will offer you, too: a chance to top off your learning in interesting and creative ways. It will ask you to think critically, to use key concepts and to answer 'big picture' questions. The capstone project is accessible online; scan the QR code to find it quickly.

Source 6

The capstone project brings together the learning and understanding in each chapter. It provides an opportunity to engage in creative and critical thinking.

Learning ladder G0.4

1 How can you use case studies to enhance your understanding of phenomena in Geography?

2 How is the How-To chapter able to assist you in your learning?

3 Suggest some classroom activities that may help you practise some of the key skills listed as part of the Learning ladder.

Changes and challenges

G1

HOW IS OUR ENVIRONMENT CHANGING?

page 14

spatial distributions and patterns

page 20

HOW DO ENVIRONMENTS CHANGE NATURALLY OVER TIME?

communicate data

page 22

HOW DO HUMANS CONTRIBUTE TO ENVIRONMENTAL CHANGE?

civics + citizenship

page 34

WHY IS ENVIRONMENTAL CHANGE A POLITICAL ISSUE?

How can I understand changes and challenges?

Our world is constantly changing. While natural factors are responsible for relatively predictable changes, human activities have resulted in a range of devastating impacts on our fragile ecosystems. Now, more than ever, we must understand the world around us and how we can change it for the better.

learning ladder

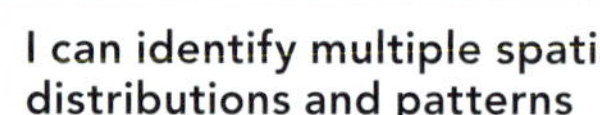

Step	Spatial distributions and patterns	Patterns and interconnections	Changes and implications
step 5	**I can identify multiple spatial distributions and patterns** I can take my PQE one step further to find links or relationships that exist in changes and challenges.	**I can interpret causes of patterns and interconnections** I can use multiple sources to find links or relationships that exist in changes and challenges and can explain 'Why?'.	**I can interpret data to quantify predictions based on research** I can use external data from research as evidence of the positive and negative impacts of a change or challenge I have predicted.
step 4	**I can use data to support exceptions to spatial distributions and patterns** I can use data to answer 'Why?' about the exceptions identified in a PQE analysis of changes and challenges.	**I can use relevant sources to research further reasons for patterns and interconnections** I can use sources other than this textbook to further research patterns I observe in changes and challenges.	**I can make predictions and outline consequences of change over time** I can use my knowledge of natural processes and world regions to make an educated guess about the positive and negative impacts of changes and challenges.
step 3	**I can describe spatial distributions and patterns** I can describe patterns, quantify them and point out exceptions (PQE) to describe changes and challenges.	**I can use data to support explanations of patterns and interconnections** I can use data from a map or graph to explain patterns I observe in changes and challenges.	**I can explain the causes behind the change over time in a place** I can use my knowledge of natural processes and world regions to explain why changes and challenges may occur over time.
step 2	**I can use data to quantify spatial distributions and patterns** I can read data and use it to measure key trends on a map or graph about changes and challenges.	**I can explain patterns and interconnections** I can identify social, historical, economic, environmental, political and technological (SHEEPT) factors to help me explain changes and challenges.	**I can describe how places have changed over time** I can use specific examples to describe changes and challenges over time.
step 1	**I can identify spatial distributions and patterns** I can find key trends on a map or graph about changes and challenges.	**I can provide short explanations for patterns and interconnections** I can write descriptions of patterns and interconnections that I find in changes and challenges.	**I can identify that changes occur in the characteristics of places over time** I can read information and answer questions about changes and challenges over time.

Source 1

Balmattum Hill on fire, threatening the town of Euroa in northeast Victoria

I can evaluate the success of research methods
I can look back and comment on the data collection methods I used, and evaluate how successful they were in helping me answer a research question about changes and challenges.

I can organise data collected according to relevance for a research question
I can review the data I have collected in the field and display it using graphs, tables, annotations and captions.

I can filter collected data
I can review my collected data and select the most relevant data to answer a research question about changes and challenges.

I can successfully use data collection methods
I can use primary and secondary data collection methods in the field and classroom to investigate changes and challenges.

I can list primary and secondary methods useful for my study
I can create a checklist of methods to investigate changes and challenges and categorise them as primary or secondary methods.

Communicate data

I can draw conclusions from geographical information in digital and geospatial technologies
I can interpret and analyse patterns by using different layers and features on geospatial technology platforms.

I can manipulate data using digital and geospatial technologies
I can work with layers and other features on geospatial technology platforms to further explore data and interconnections.

I can access and use geospatial technology platforms such as GIS
I can use geospatial technology platforms to explore data and find patterns.

I can construct paper maps using correct cartographic conventions
I can use a pencil, paper and ruler to construct a map that follows BOLTSS conventions.

I can interpret different map types using cartographic conventions
I understand data found in different types of maps and graphs and use the data to answer questions about changes and challenges.

Digital and geospatial technologies

Warm up

Spatial distributions and patterns

1 Look around the classroom and describe the *distribution* of the desks.

Patterns and interconnections

2 Describe the *interconnection* between changing global climate and the occurrence of bushfires.

Changes and implications

3 Source 1: Research some of the outcomes of the 2019–2020 bushfires in Victoria. Outline how *places* were *changed* and the implications of these changes.

Communicate data

4 Source 1: Imagine you were doing fieldwork in this location to collect data on what resources locals needed to rebuild and recover. Suggest two primary methods that may be useful in this project.

Digital and geospatial technologies

5 Suggest how digital mapping may have been used by different stakeholders during the 2019–2020 bushfire season.

How is our environment changing?

All environments undergo change over varying time scales. Many changes are a result of natural processes such as the movement of tectonic plates, shifts in trade winds and oceanic currents, or as a result of natural disasters. However, as the human population continues to grow, our rapid **urbanisation**, agriculture expansion and reliance on industry means we are responsible for significant **terrestrial**, **marine** and **atmospheric** changes.

Source 1

These satellite photos show change over time: **a** early spring in the Kulunda Steppe, on the Russia-Kazakhstan border, 1 April 2005 and **b** late spring in the Kulunda Steppe, 11 June 2005.

Global human modification on terrestrial systems

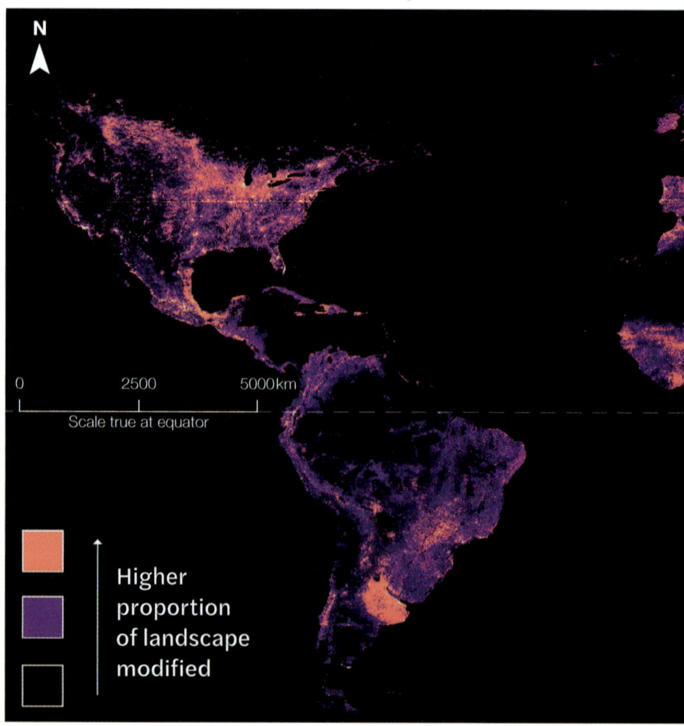

Source 2

This heat map shows human impact on Earth.

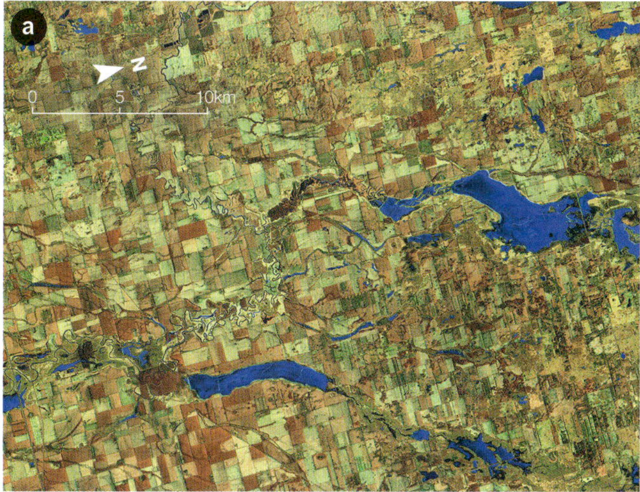

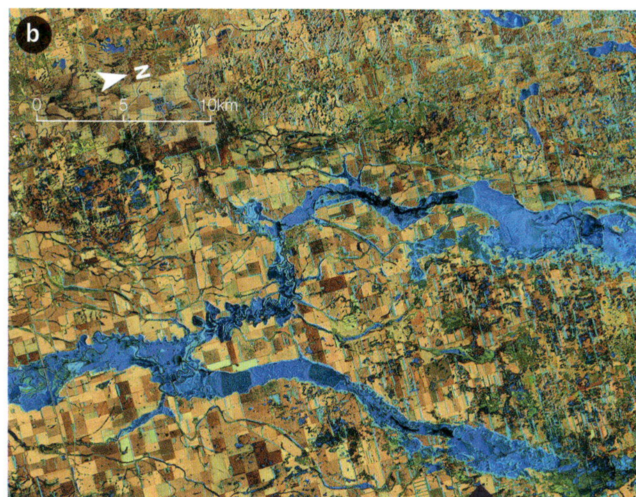

Source 3

These satellite photos show change over time: **a** the James River, in South Dakota, USA, 10 March 2015 and **b** the James River flooding, 23 March 2020.

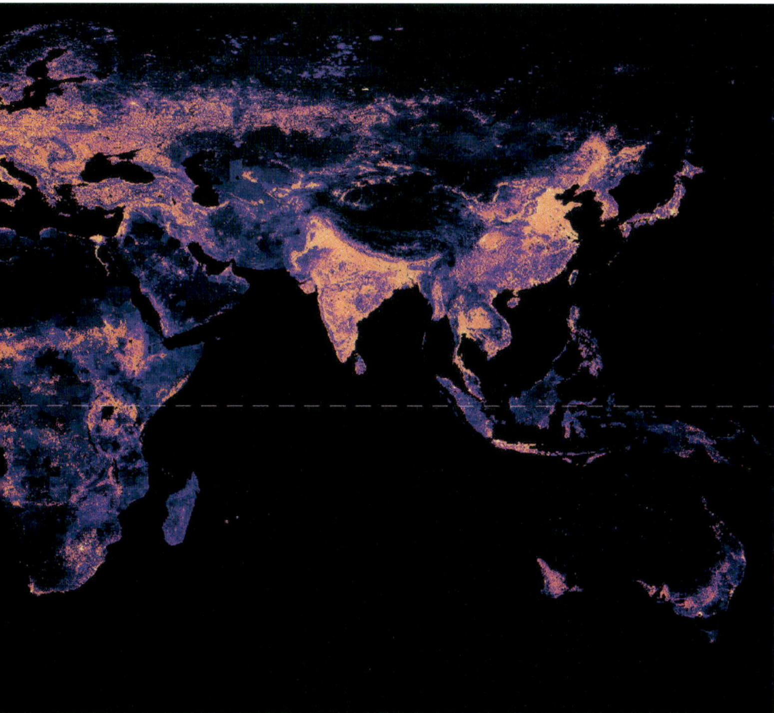

Source: Hannah Ker

Source 4

The Pedersen Glacier in Alaska. These photos show significant change over time because of climate change.

Learning ladder G1.1

Show what you know

1 Suggest one natural factor and one human factor that may lead to changes in an *environment*.

2 Using the correct conventions, complete a sketch that illustrates some environmental change that has occurred on a local *scale* in your *region*.

Changes and implications

Step 1: I can identify that changes occur in the characteristics of places over time

3 Select a source from this spread and identify two *changes* that have occurred over time.

Step 2: I can describe how places have changed over time

4 Compare two of the sources on this spread. Suggest which *region* has seen more *change* by outlining the characteristics of each location.

Step 3: I can explain the causes behind the change over time in a place

5 Select one location from this spread and conduct further research on the *changes* that have occurred there. Are these changes largely attributed to natural factors or human activity? What evidence helped you form this conclusion?

Step 4: I can make predictions and outline consequences of change over time

6 'All environmental change is negative'. Discuss with reference to at least one example.

Sketches and annotating, page 158
Answering questions, page 162

What are the effects of the rising demand for land?

Global population growth over the past century has been nothing short of remarkable. From a total population of around 2 billion in 1920, our population has almost quadrupled to around 7.6 billion in just 100 years. This has coincided with a shift in lifestyles and land usage; as our cities grow bigger, so does our demand for arable land.

The majority of the world's urban areas are found in arable grassland and forest biome regions. As cities expand, it often means that farmland and market gardens on the peri-urban fringe are increasingly converted to residential, commercial and industrial areas. As populations continue to grow, this can increase or exacerbate the risk of food insecurity, and push farmers to use increasingly marginal land.

Source 1

Egyptian farmland close to the Nile

Arable land use per person, 2015

Arable land is defined as land under temporary crops, temporary meadows for mowing or for pasture

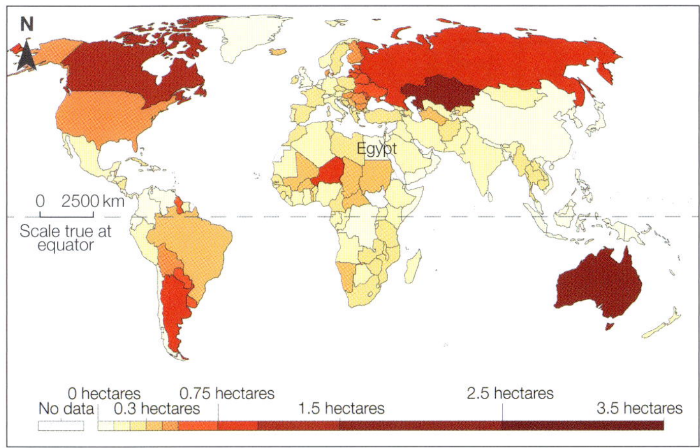

Source: Our World in Data

Population density, 2015

The number of people per km^2 of land area

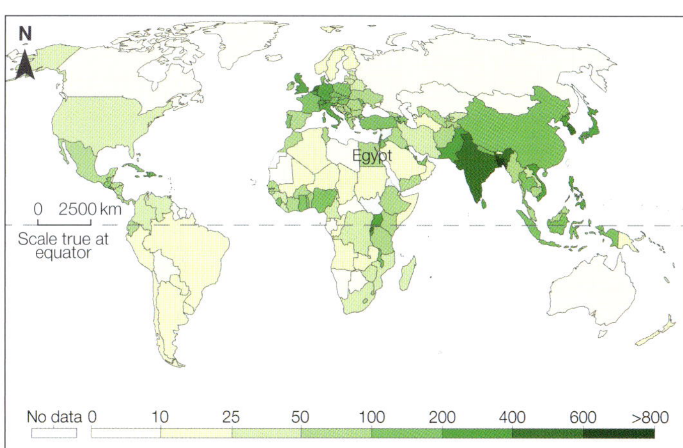

Source: Our World in Data

Source 2

These world maps show arable land use in 2015, measured in hectares per person, and the population density of Earth in the same year.

Only 2.8 per cent of Egypt's land is considered arable, primarily the narrow areas adjacent to the Nile River and its delta. In a 2017 study, researchers predicted that, as population growth continues, agricultural yield will fall: maize production is predicted to fall by 60 per cent by 2030 while rice could fall by 40 per cent.

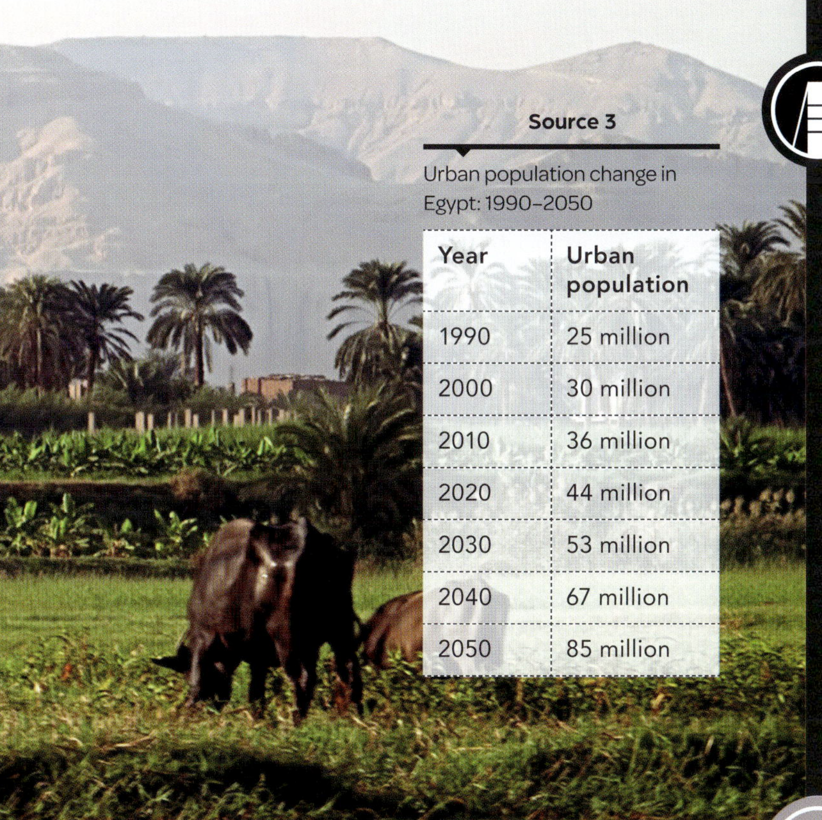

Source 3

Urban population change in Egypt: 1990–2050

Year	Urban population
1990	25 million
2000	30 million
2010	36 million
2020	44 million
2030	53 million
2040	67 million
2050	85 million

Learning ladder G1.2

Show what you know

1 Outline how the global population has *changed* over the past century.

2 Suggest why agricultural land on the peri-urban fringe is being 'lost'.

Changes and implications

Step 1: I can identify that changes occur in the characteristics of places over time

3 Source 3: describe how Egypt's urban population has *changed* over time.

Step 2: I can describe how places have changed over time

4 What is happening to some areas of arable land in Egypt? Is a similar pattern likely in Melbourne?

Step 3: I can explain the causes behind the change over time in a place

5 Egypt's ability to produce food is falling; at the same time its population is increasing. With reference to *process*, explain how this is possible.

Step 4: I can make predictions and outline consequences of change over time

6 'Urban growth can actually lead to increased food prices and land degradation.' To what extent do you agree with this statement? Prepare a mind map to illustrate how this *change* can occur.

HOW TO

PQE, page 154
Answering questions, page 162

How can we use geospatial technology to measure environmental change?

GPS, Google Maps and satellite imagery are all examples of geospatial technologies. Geospatial technology allows geographers and all of us to visualise patterns on a variety of physical and temporal scales. Digital maps display data in layers over a base map. By putting multiple layers of data on a map, we are able to overlay information and better observe spatial associations and interconnected processes. Geographers are able to then use these visualisation tools to assess, predict and manage environmental changes.

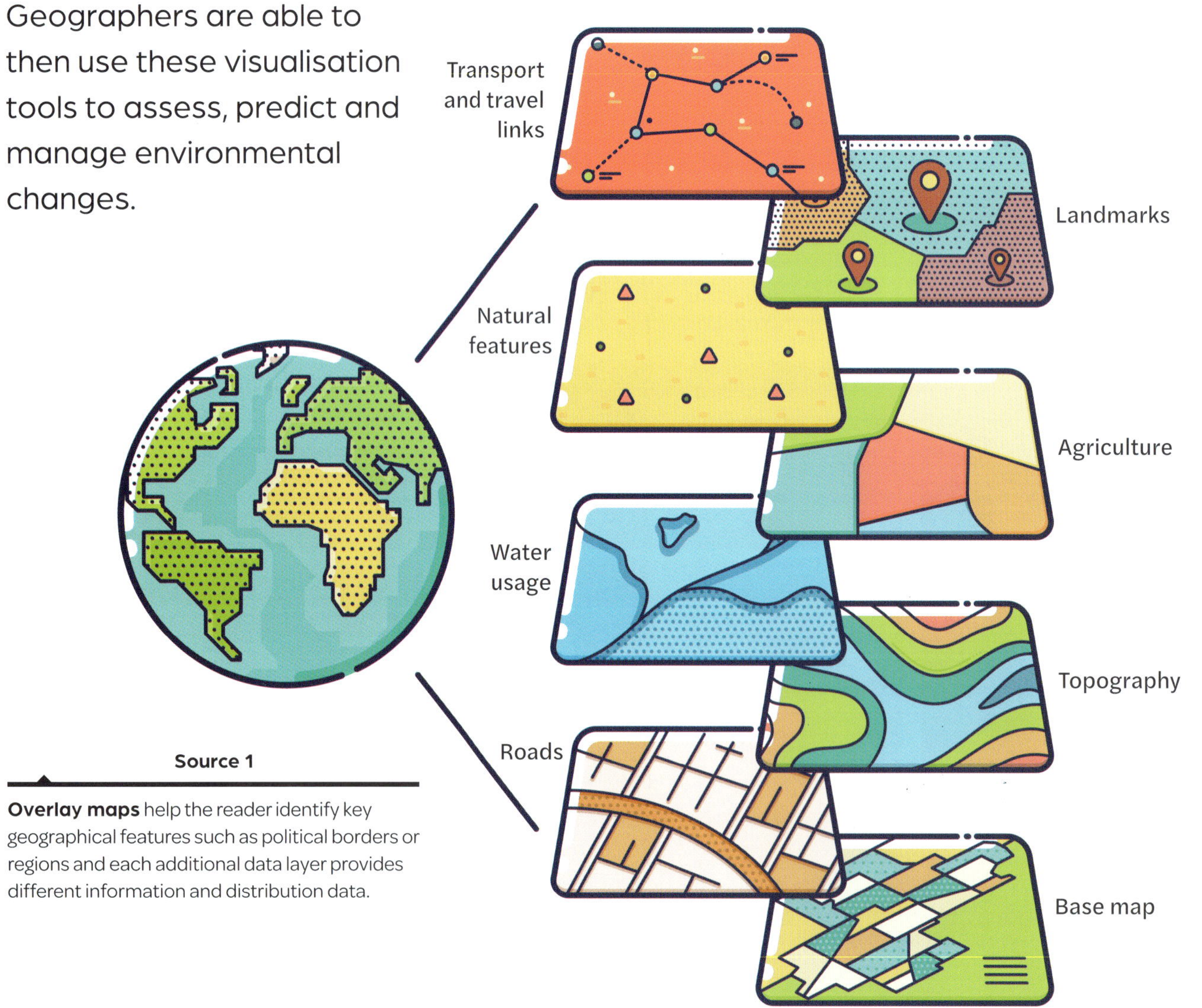

Source 1

Overlay maps help the reader identify key geographical features such as political borders or regions and each additional data layer provides different information and distribution data.

Degrees cooler or warmer in 2020 compared to the middle of the 20th century

Canada
UnitedStates of America
Russia
China
Pacific Ocean
Atlantic Ocean
South Africa
Indian Ocean
Australia
Southern Ocean
Antarctica

No data | -2°C | -1°C | -0.5°C | -0.2°C | 0°C | 0.2°C | 0.5°C | 1°C | 2°C | 4°C

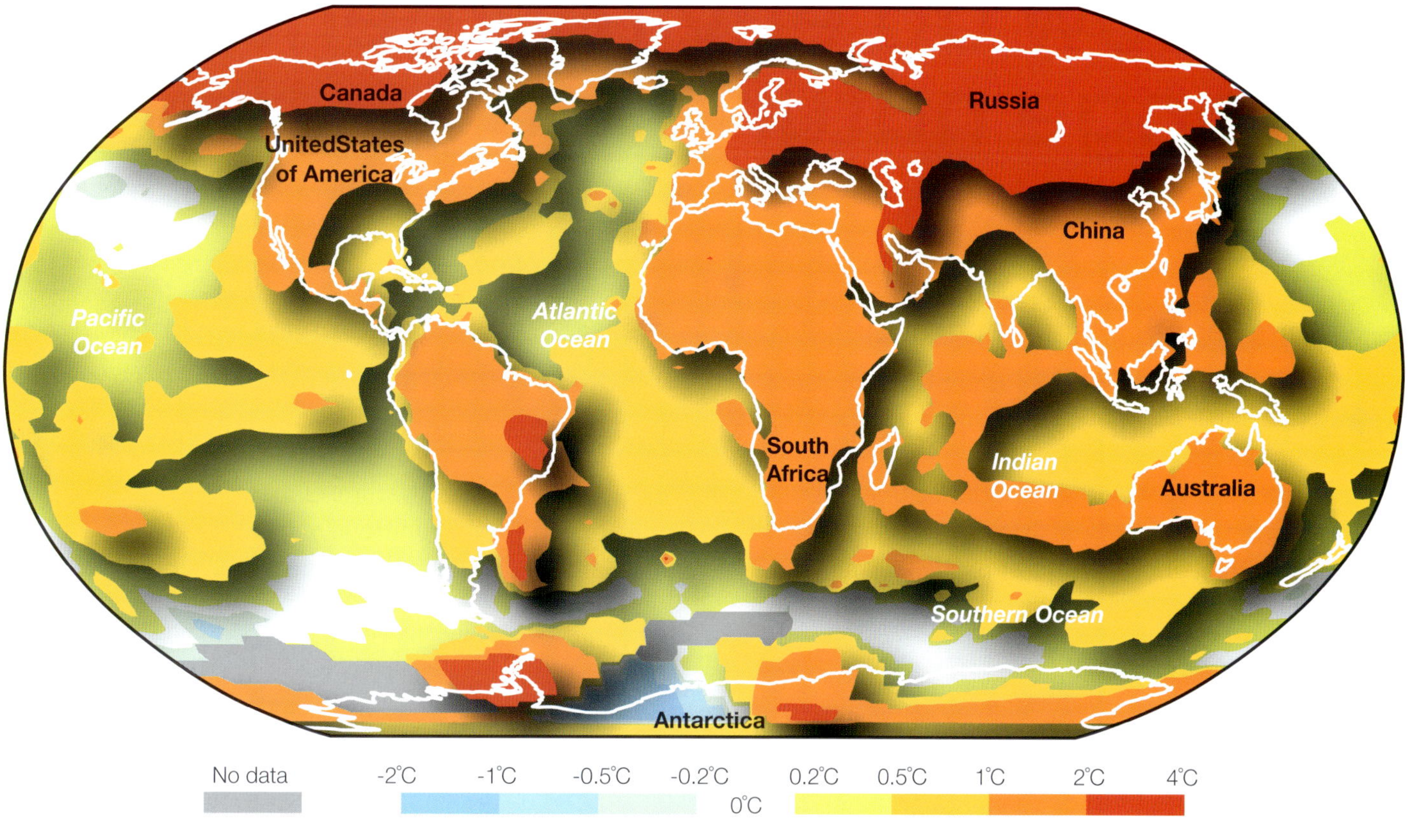

Source 2

The data layer over this political map shows temperature changes on a global scale in 2019 compared to those in the middle of the 20th century.

Source: Matilda Education Australia/ Custom Mapping Services

Learning ladder G1.3

Show what you know

1 Define the term 'geospatial technology'.

2 Suggest two ways geospatial technology differs from other forms of technology.

3 List three ways we can use mapping technology to track or manage environmental *change*.

Digital and geospatial technologies

Step 1: I can interpret different map types using cartographic conventions

4 Source 2: Outline the data that is displayed on this map and how it is *interconnected* with environmental *change*.

Step 2: I can construct paper maps using correct cartographic conventions

5 a Using BOLTSS, locate the following glaciers on a world map: Fedchenko Glacier, Siachen Glacier, Biafo Glacier, Bruggen Glacier and South Inylchek Glacier.

b Compare the locations of these glaciers to the warming temperatures in Source 2. Predict how these glacial environments may *change* in the future.

Step 3: I can access and use geospatial technology platforms such as GIS

6 Access http://mea.digital/GHV10_G1_1 and log in as a 'guest'. List the data that can be overlaid on the base map.

Step 4: I can manipulate data using digital and geospatial technologies

7 Access http://mea.digital/GHV10_G1_1 again. Select one country and, by turning on and off layers on the data panel, explore the environmental *change* that is occurring.

Mapping with BOLTSS, page 152
PQE, page 154
Answering questions, page 162

How do environments change naturally over time?

Natural changes tend to occur over relatively predictable or long time periods and are interconnected to help sustain environments and **biodiversity**. Annual or seasonal changes bring about shifts to temperature, rainfall, vegetation density or snow pack.

For example, in autumn, as deciduous trees lose their leaves, the fallen leaf litter acts as compost, improving soil quality. In the spring, snow melts and rainfall increases to form rivers and streams, moving fresh water and nutrients to new areas. Animal species rely on these environmental changes for food and to time important events such as migrations or mating seasons (Source 1).

Long-term natural changes such as those we see in the global climate also show relatively consistent patterns and cycles. While the terms **'climate change'** and **'global warming'** are often used interchangeably, the term 'climate change' more accurately describes the global average temperature fluctuation over time due to a range of natural processes. Natural climate change has been occurring for hundreds of thousands of years and is responsible for the ice ages and warm periods (Source 2).

Source 1

A 20-year study published by the CSIRO found that sea surface temperature and ocean currents help little penguins in Western Australia time their breeding season so they can have the most successful fledgling rate.

Natural climate change is strongly interconnected with another natural process known as the greenhouse effect. The Earth's atmosphere contains various greenhouse gases such as carbon dioxide. As the Sun's energy radiates downward, some of the Sun's heat is absorbed by Earth's surface, while the rest is reflected back into the atmosphere. Greenhouse gases act as a blanket capturing some reflected heat so the Earth stays at a habitable temperature (Source 3). When there is more carbon dioxide in the atmosphere, the global climate warms. As a result, carbon dioxide levels are often used to show natural fluctuations in climate and therefore environmental changes over long periods.

Carbon dioxide levels over 800 000 years

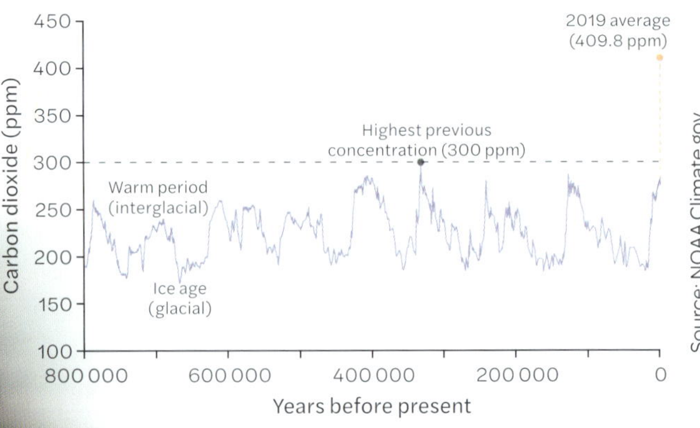

Source 2

Carbon dioxide levels in the atmosphere have been changing naturally over time and have resulted in ice ages and warm periods (interglacial periods).

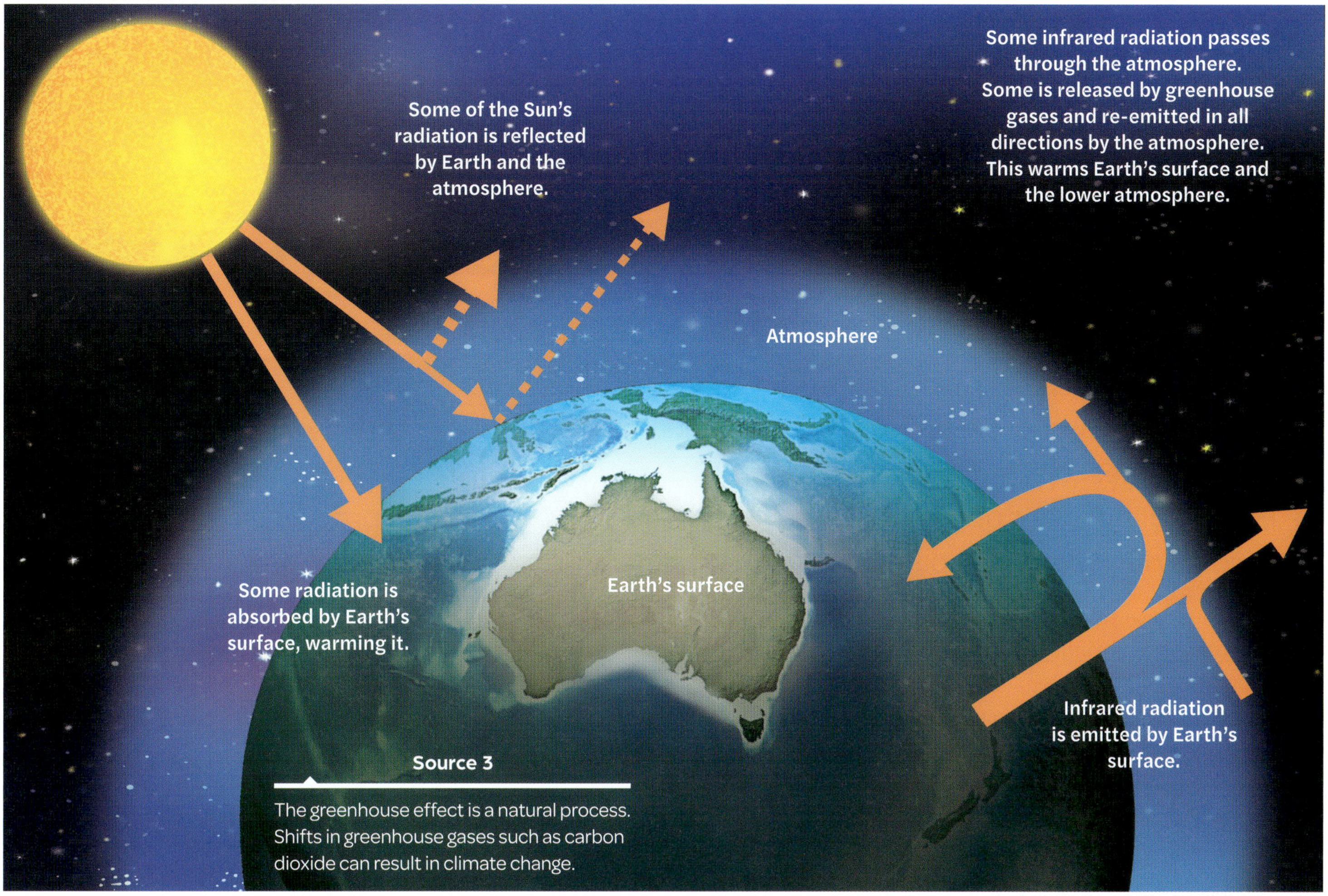

Some of the Sun's radiation is reflected by Earth and the atmosphere.

Some infrared radiation passes through the atmosphere. Some is released by greenhouse gases and re-emitted in all directions by the atmosphere. This warms Earth's surface and the lower atmosphere.

Atmosphere

Earth's surface

Some radiation is absorbed by Earth's surface, warming it.

Infrared radiation is emitted by Earth's surface.

Source 3

The greenhouse effect is a natural process. Shifts in greenhouse gases such as carbon dioxide can result in climate change.

Learning ladder G1.4

Show what you know

1 Outline why natural *change* is essential for ecosystems.

2 Create a cartoon or schematic highlighting the *process* of the greenhouse effect.

3 Complete a sketch of a local area that highlights key environmental changes that have occurred in your lifetime.

Spatial distributions and patterns

Step 1: I can identify spatial distributions and patterns

4 Source 2: Describe the pattern of atmospheric carbon dioxide levels over time.

Step 2: I can use data to quantify spatial distributions and patterns

5 Source 2: Compare the highest previously recorded carbon dioxide concentration with that recorded in 2019.

Step 3: I can describe spatial distributions and patterns

6 Using PQE, describe the *changes* that have occurred in carbon dioxide concentration from 200 000 years ago until 2019.

Step 4: I can use data to support exceptions to spatial distributions and patterns

7 Using information from this section, suggest why global warming sceptics may see our human-induced warming as part of a natural pattern.

PQE, page 154
Sketches and annotating, page 158
Answering questions, page 162

How do humans contribute to environmental change?

Environmental change brought about by humans generally occurs over a much shorter timescale, is more extensive and less than naturally occurring changes. Human activity not only changes environments directly through processes such as **deforestation**, but can also alter longer-term cycles such as climate.

Unlike naturally occurring climate change, global warming is a human-induced process where we have dramatically increased the carbon dioxide in the atmosphere leading to an **enhanced greenhouse effect** and the unnatural, **exponential** warming of the globe. Recent studies have revealed that carbon dioxide in the atmosphere is the highest it has been in the past 400 000 years, leading to an average global temperature increase of 0.8°C since 1880. These increased global temperatures are responsible for melting glaciers and sea ice, increasing the occurrence of extreme weather events, reducing predictability of growing seasons for crops and loss of biodiversity on a global scale.

Source 1

As the levels of carbon dioxide and other greenhouse gases increase in our atmosphere, more solar radiation is trapped leading to global warming.

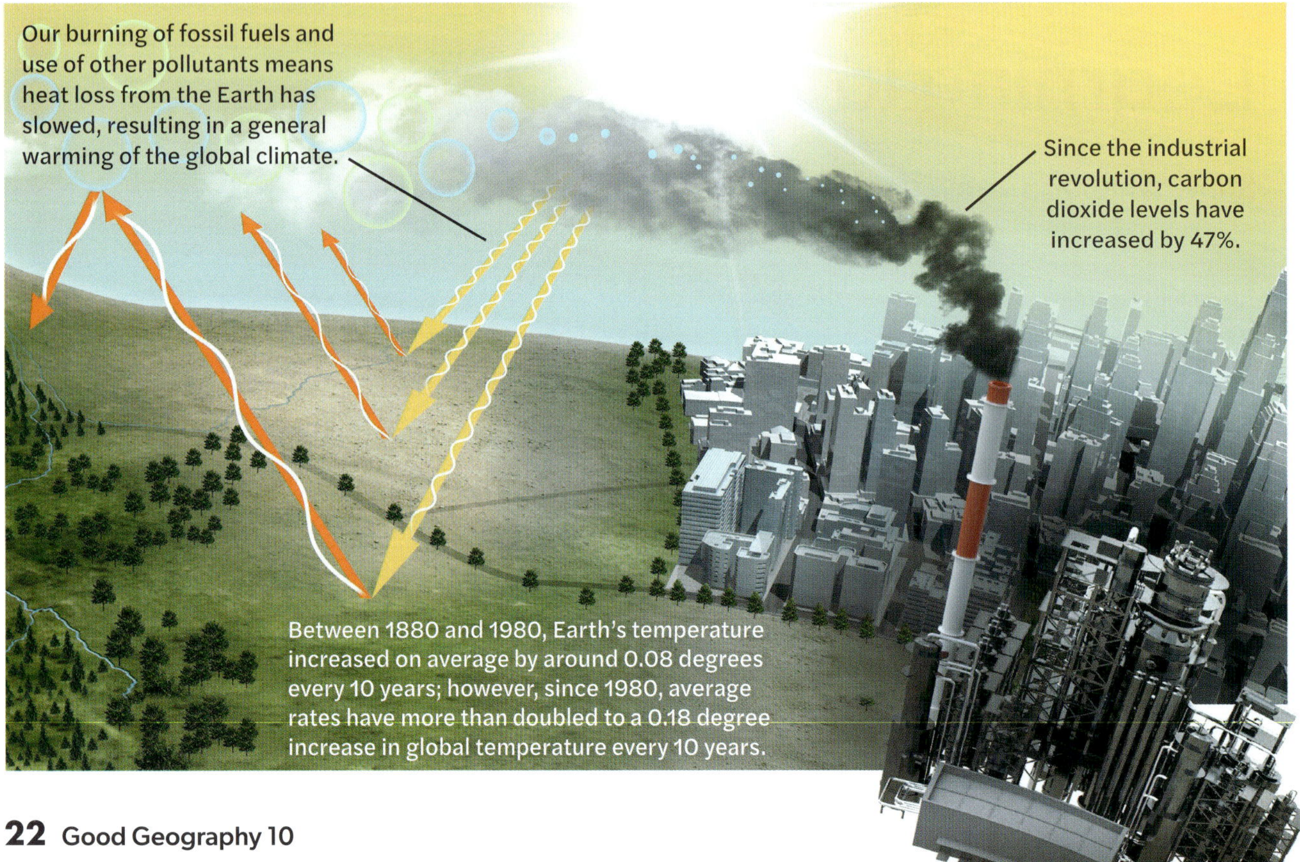

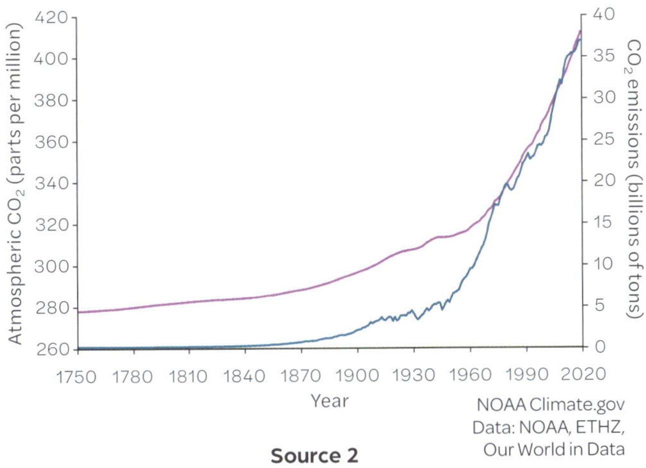

Source 2

There is a clear correlation between our emission output (the blue line on the graph) and the level of carbon dioxide present in the atmosphere (the pink line). Atmospheric greenhouse gas levels have increased exponentially since the industrial revolution and are not consistent with natural patterns.

Source 3

Human activities lead to a variety of large-scale and dramatic environmental changes such as **a** deforestation for agriculture and **b** urbanisation.

Learning ladder G1.5

Show what you know

1 List three ways humans alter *environments*.

2 Describe the difference between the natural greenhouse effect and human-enhanced greenhouse effect.

3 Why are human *changes* to environments less predictable than natural changes?

Communicate data

Step 1: I can list primary and secondary methods useful for my study

4 List two secondary websites that are reliable for collecting data on the consequences of human-induced environmental *change*.

Step 2: I can successfully use data collection methods

5 Create a photo essay that highlights some environmental *change* that is occurring on a global *scale* as a result of human activity.

Step 3: I can filter collected data

6 Look over chapters G1 and G2 and identify two quantitative data sources and two qualitative data sources.

7 Using the correct conventions, create a sketch map that highlights key places in your local *region* affected by human activity over time. Use a suitable legend to show whether these changes have been positive or negative for the natural *environment*.

Step 4: I can organise data collected according to relevance for a research question

8 You are working as part of a student consultant group for your local council that is trying to reduce the impact human activity has on local environments. Suggest two long-term and two short-term approaches for the council to consider.

Sketches and annotating, page 158
Photo essays, page 160
Answering questions, page 162

How do we track environmental change over time?

Our understanding of long-term climate cycles and changes to environments over time is helping us understand the impact of human activities on our planet. Geographic data plays a critical role in determining natural change and fluctuation, as well as understanding past changes and predicting likely future changes.

Geospatial technologies, primarily the use of satellite imaging technology and geographic information systems (GIS), allow us to directly compare areas on different scales, over time. The possibilities for tracking human activities and natural processes are unlimited, from deforestation to desertification, from urban growth to the movement of migratory species. Data sources, including those from citizen science programs, can further enhance geospatial technology and the visualisation of information.

Ice core sampling (Source 1) is another method for tracking climatic change, using samples drawn from ice sheets and glaciers to obtain hydrogen and oxygen isotopes trapped within them. These reveal information about temperature changes and the levels of atmospheric gases. To date, ice cores from Vostok Station, Antarctica have been drawn to a depth of more than 3.7 km, revealing almost 800 000 years of climate change data. This supports projections, based on temperatures and gas levels, of global land cover during periods of glacial maximums and climatic optimums.

Source 1

Ice core sampling is an important measure of historical climate data.

First Nations' perspectives on change over time

According to the World Bank, there are more than 475 million indigenous peoples in the world today, yet indigenous environmental understanding remains underutilised. In Australia, First Nations' knowledge has provided valuable insights into the seasonal cycles of resources, landscapes and ecology, honed through tens of thousands of years of tradition and a relationship of respect with the natural world. Different language groups around Australia have unique seasonal calendars to mark key changes in flora, fauna and climate. For example, the Gulumoerrgin (Larrakia) of Darwin and the surrounding regions mark seven different seasons throughout the year.

Understanding the seasons and the rich traditional knowledge intertwined with language is a core part of tracking environmental change over time.

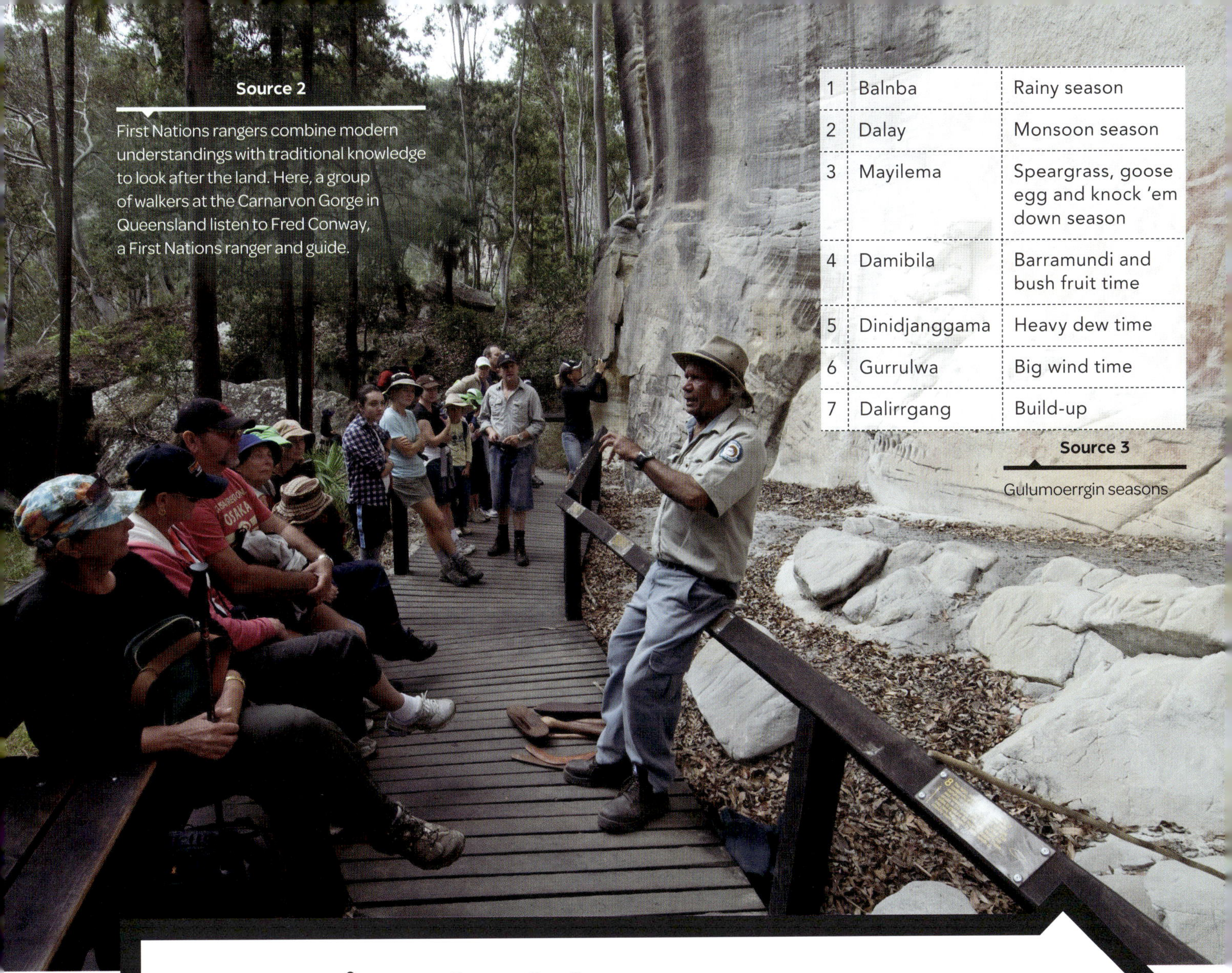

Source 2

First Nations rangers combine modern understandings with traditional knowledge to look after the land. Here, a group of walkers at the Carnarvon Gorge in Queensland listen to Fred Conway, a First Nations ranger and guide.

1	Balnba	Rainy season
2	Dalay	Monsoon season
3	Mayilema	Speargrass, goose egg and knock 'em down season
4	Damibila	Barramundi and bush fruit time
5	Dinidjanggama	Heavy dew time
6	Gurrulwa	Big wind time
7	Dalirrgang	Build-up

Source 3

Gulumoerrgin seasons

Learning ladder G1.6

Show what you know

1 What are the three main ways we can track environmental *change* over time?

2 Why might some First Nations people recognise more than four seasons?

Changes and implications

Step 1: I can identify that changes occur in the characteristics of places over time

3 Outline why ice core sampling is a valuable resource for understanding climate change over time.

Step 2: I can describe how places have changed over time

4 With reference to a specific location or *region*, describe how satellite images of the same area over time might be useful for environmental management.

Step 3: I can explain the causes behind the change over time in a place

5 On Crocodile Island, in north-east Arnhem Land, traditional owners work as rangers to care for the land and educate the next generation. Under their stewardship, they successfully reduced pest species, such as the cane toad, while preserving traditional flora and fauna. Explain why they have been more successful than other rangers.

Step 4: I can make predictions and outline consequences of change over time

6 Traditional knowledge, in Australia and many other parts of the world, contains a special understanding of the relationship between people and the *environment*. Why has such knowledge sometimes been misunderstood or ignored by decision-making bodies? How could this knowledge improve the ways environmental *change* is tracked?

Answering questions, page 162 HOW TO

What are the challenges of sustainability?

While **sustainability** is usually associated with environmental longevity, as geographers we also need to consider economic and societal sustainability. That is, maintaining a viable economy and creating an equal and fair society for future generations. Often there is a conflict between what is best for economic and environmental sustainability.

Forests are vital for many of Earth's key processes. Trees absorb carbon dioxide and release oxygen and so provide one of the most successful natural ways to reverse the impacts of global warming. Large forests are also linked to atmospheric circulation and global precipitation levels, and are home to over 80 per cent of our terrestrial plant and animal species.

The timber industry, however, employs over 54.2 million people worldwide and contributes over $600 billion annually to the global economy. Further, land clearing for logging makes more land available for large-scale agriculture. Many people, especially those in **less economically developed countries (LEDCs)** rely on such industries for their livelihood.

Environmentalists who push for a reduction in deforestation and other environmental changes can come into conflict with those who focus on the potential monetary loss for communities and countries. As geographers, we need to see both sides of the issue and suggest solutions that have benefits for all stakeholders.

Source 1

Deforestation is an example of a conflict between economic and environmental sustainability.

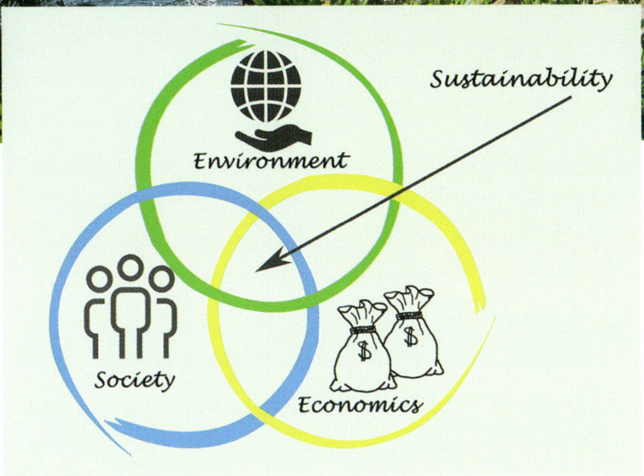

Sustainability
Environment
Society
Economics

Source 2

Sustainability is about more than just the environment. We need to ensure that we are also sustainable in our economy and society to ensure a peaceful and safe place for future generations.

Source 3

Barangaroo Reserve is a regeneration project in Sydney, NSW. Human-induced environmental change is not always negative. Humans can positively impact places through regeneration projects, creating protected areas and reducing overfishing or over farming. Unfortunately, many of our positive changes are simply correcting human-caused damage of the past.

Learning ladder G1.7

Show what you know

1 Explain the *interconnection* between environmental, economic and societal *sustainability*.

2 Outline how human activity can lead to conflict about managing the *environment*.

Patterns and interconnections

Step 1: I can provide short explanations for patterns and interconnections

3 Suggest another example of a conflict of *sustainability* between economists and environmentalists. Explain why this conflict exists.

Step 2: I can explain patterns and interconnections

4 Access http://mea.digital/GHV10_G1_2. Watch the *change* over time in world forest *distribution* from 1990–2017. Explain why forest cover may have reduced in some *regions* over this period.

Step 3: I can use data to support explanations of patterns and interconnections

5 Access http://mea.digital/GHV10_G1_2 again. Watch the *change* over time in world forest *distribution* from 1990–2017. Using data, comment on the increase in forest cover in one country. Provide two reasons why forest cover may be increasing in this *place*.

Step 4: I can use relevant sources to research further reasons for patterns and interconnections

6 Consider the following people's views, wealth or position and comment on how their views might influence local environmental management:

a a low-income worker relying on timber milling for a basic wage

b an environmentalist who manages orphaned or abandoned wildlife

c a politician whose country earns most of its **Gross Domestic Product (GDP)** from timber- and forestry-related products.

PQE, page 154
SHEEPT, page 156
Answering questions, page 162

How does urbanisation change the environment?

Climate change data shows how, over the last 800 000 years, our Earth has cycled between periods of glacial maximums and climatic optimums, affecting temperature and precipitation levels and greatly impacting natural land cover.

Since the Industrial Revolution, this cycle of environmental change has greatly accelerated, primarily as a result of urbanisation. In 1800, the global population reached one billion for the first time and, by 1920, it had doubled to two billion.

A century later, the population has reached almost 7.7 billion people and economies have shifted drastically from subsistence agriculture to manufacturing and commercial agriculture. Interconnections and the movement of people and goods between regions has increased, as has the infrastructure to support this flow; in terms of environment change, this has had an enormous impact. Population growth and industrialisation have led to:

- the demand for more land to expand urban areas, often resulting in the loss of arable land. Urban land areas currently expand twice as fast as urban populations, especially across the peri-urban fringe
- the loss of biodiversity as forest and grassland biomes are cleared for agriculture and urban growth. It is estimated that urban growth accounts for more than seven million hectares of arable land per year
- an increase in the release of greenhouse gases through industrialisation
- the loss of soil nutrients as food production increasingly shifts from subsistence farming to commercial agriculture. The use of pesticides and other chemicals to increase yield also accelerates land and water degradation
- a rise in the amount of waste and landfill that needs to be disposed of; at present, about 2 billion tonnes of waste is produced each year
- widespread breeding of animals for food, including cows, sheep and pigs; this consumes more than 10 million acres of arable land per year
- increased surface temperatures in cities due to the urban heat island effect, where increased reflection and reduced vegetation create sweltering conditions – see Source 2.

Projected waste generation, by region (millions of tonnes/year)

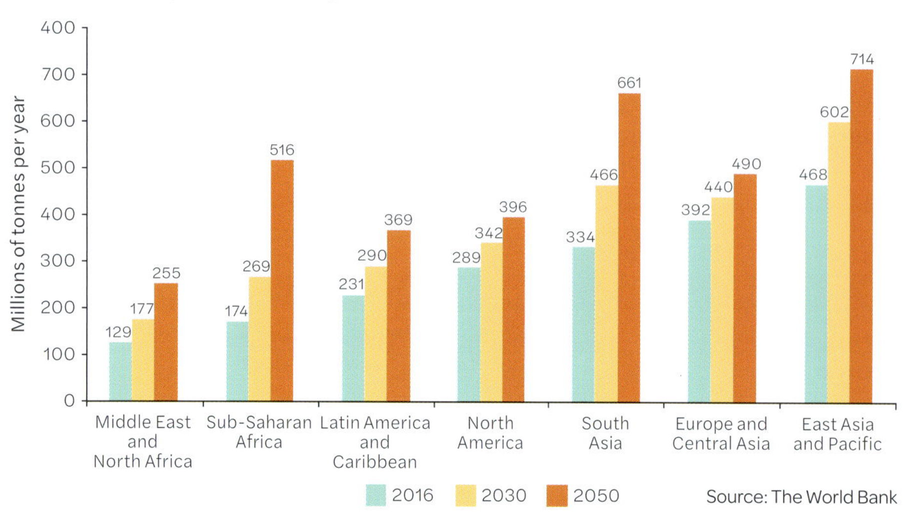

Source 1

Ever-increasing waste is an ongoing environmental problem.

Source 2

The growth of urban areas has led to a decrease in the presence of natural elements that help cool the air temperature. This is called the urban heat island effect.

Sunlight

Temperature increase

Heat of vaporisation

Heat from the bulding surface

Temperature goes down

Heat of vaporisation

Temperature increase

Temperature increase

Temperature goes down

Anthropogenic heat

Heat from vehicles

Air conditioners

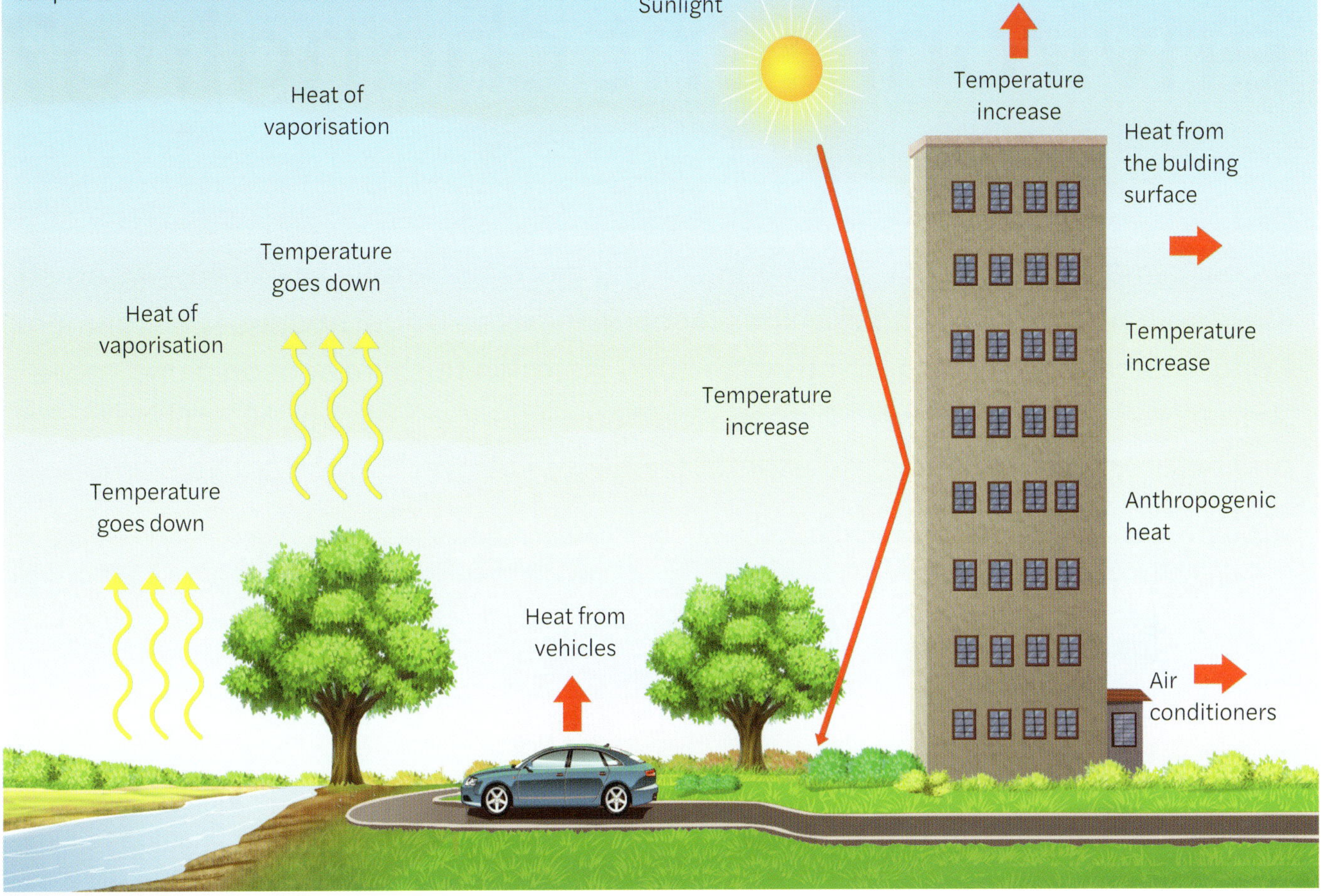

Learning ladder G1.8

Show what you know

1 Define the urban heat island effect.

2 How has our global population *changed* over the past two centuries? What does this mean for cities and farms?

Changes and implications

Step 1: I can identify that changes occur in the characteristics of places over time

3 Identify three negative ways that the growth of cities can impact the *environment*.

Step 2: I can describe how places have changed over time

4 Source 1: Describe how waste generation around the world could change from 2016 to 2050.

Step 3: I can explain the causes behind the change over time in a place

5 Explain what causes the urban heat island effect and suggest how it might affect the local *environment*.

Step 4: I can make predictions and outline consequences of change over time

6 'Agriculture, industrialisation and urban growth can all be balanced to ensure a healthy planet.' Using examples and *interconnection*, evaluate this statement.

PQE, page 154
SHEEPT, page 156
Answering questions, page 162

How is Australia's environment changing?

As our climate changes, Australia is becoming more vulnerable to extreme weather events such as drought and bushfires. According to the Department of Environment, Land, Water and Planning, in the 2019–2020 Victorian bushfire season alone, more than 1.2 million hectares were burned including 60 per cent of over 50 national parks and nature reserves, leading to estimates of a loss of over 1 billion native animals.

While drought and other climate events contributed to the severity of these fires, there is also an interconnection with local land cover changes. While it may be assumed that logging and large-scale forest clearance could reduce our risk of fire, Australian scientists are warning that removal of native old growth forests could lead to an increase in fire intensity over time. Large-scale logging means that areas that were once dense, wet, tropical forests are altered into largely open shrub and dry bushland. This reduces soil moisture and exposes the small number of trees to wind and potential **ember** attack. Further, the act of logging leaves more debris on the ground to act as fire fuel.

Large forested areas are also vital for maintaining our national climate. It has been estimated that, in eastern Australia, our summer surface temperatures have increased over 2 degrees as a result of reduced tree coverage. While the removal of some forests may be necessary for urbanisation, agriculture and economic stability, we need to find a balance in order to also have environmental and climatic sustainability.

Temperature change °C

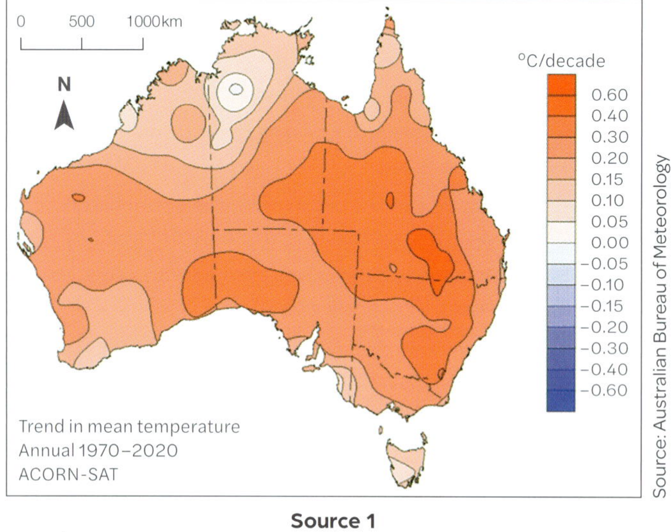

Source 1

According to the CSIRO, Australia's mean temperature has warmed by around 1°C since 1910. This map shows the temperature increases from 1970.

Change in number of dangerous fire weather days

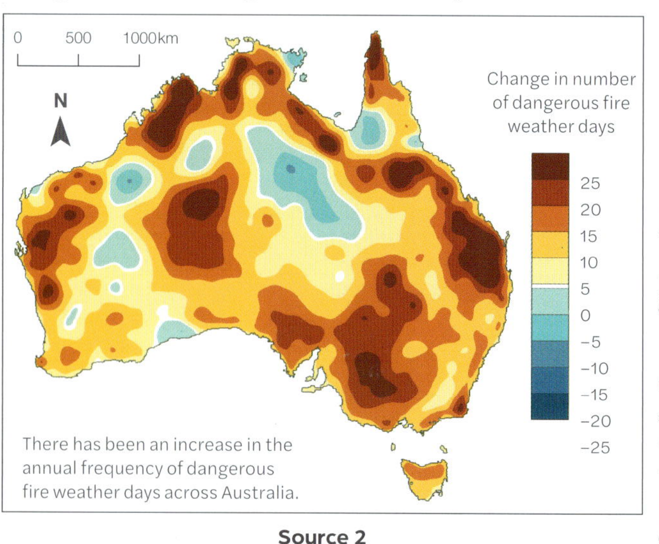

Source 2

Fire weather conditions are mostly worsening, particularly in the south and east.

Average annual temperatures are predicted to increase

More extreme weather events predicted, resulting in more severe drought and flood

Snow pack is likely to decrease in mountain regions, meaning a loss in recreation, habitat and seasonal albedo

Monsoon rain patterns in northern Australia predicted to change, affecting natural cycles

Fire seasons longer and more severe

Ocean temperatures will continue to increase, causing increased acidity, bleaching of coral reefs and unpredictability of natural patterns.

Sea levels predicted to increase and inundate low lying regions, including some major cities

Soil moisture is predicted to decrease as a result of increased temperatures meaning there is more surface evaporation. This will mean it is more difficult for plants to grow and survive

Source 3

The effects of climate change in Australia

Learning ladder G1.9

Show what you know

1 List some types of environmental *change* occurring in Australia.

2 Summarise how human activity increases the risk of extreme weather events in Australia.

3 Create a mind map that illustrates some challenges to *sustainability* in Australia.

Spatial distributions and patterns

Step 1: I can identify spatial distributions and patterns

4 Create a flow diagram to illustrate how increased temperature is connected with fire danger in Australia.

Step 2: I can use data to quantify spatial distributions and patterns

5 Source 1: Divide the class into small groups and allocate a state or territory to each. Using data, outline how temperature has changed in this place.

Step 3: I can describe spatial distributions and patterns

6 Describe the *spatial association* between increased temperature and fire severity in Australia.

Step 4: I can use data to support exceptions to spatial distributions and patterns

7 Source 3: Some locations in Australia do not have an increased risk of fires in the future. Describe the locations of these *places* and use data to support this statement.

PQE, page 154
SHEEPT, page 156
Answering questions, page 162

How can global changes have local impacts?

As Earth's climate continues to warm over time, we will begin to observe more extreme environmental changes on a variety of scales. One impact that is widely discussed is the melting of glaciers and ice sheets.

Ice and snow have a high **albedo**, meaning that up to 90 per cent of the Sun's energy is reflected, keeping the region and global climate cooler. As this ice melts, less of the Sun's energy is reflected. Melt water causes global sea level rise and wider expanses of dark ocean result in more atmospheric heat being absorbed, furthering global warming.

Source 3 is one of the most iconic and emotive images representing the impacts of global warming on polar environments. But how will this melting affect us long term? The National Oceanic and Atmospheric Administration (NOAA) recently reported that sea level rise has more than doubled, and, in 2019, 'global mean sea level was 87.61 mm above the 1993 average—the highest annual average in the satellite record (1993-present)'.

Many of our world's largest and most populated cities are located in coastal regions. Around 85 per cent of Australians live within 50 km of the coast. As sea levels rise, these places will be vulnerable to stronger storm surges and increased flooding, which threatens infrastructure, trade routes and economic stability. Floods can also inundate waste removal areas, which could then pollute local waterways. Australia may lose its iconic beaches, coral reefs and wetlands meaning a loss in biodiversity as well as a potential loss in tourism earnings.

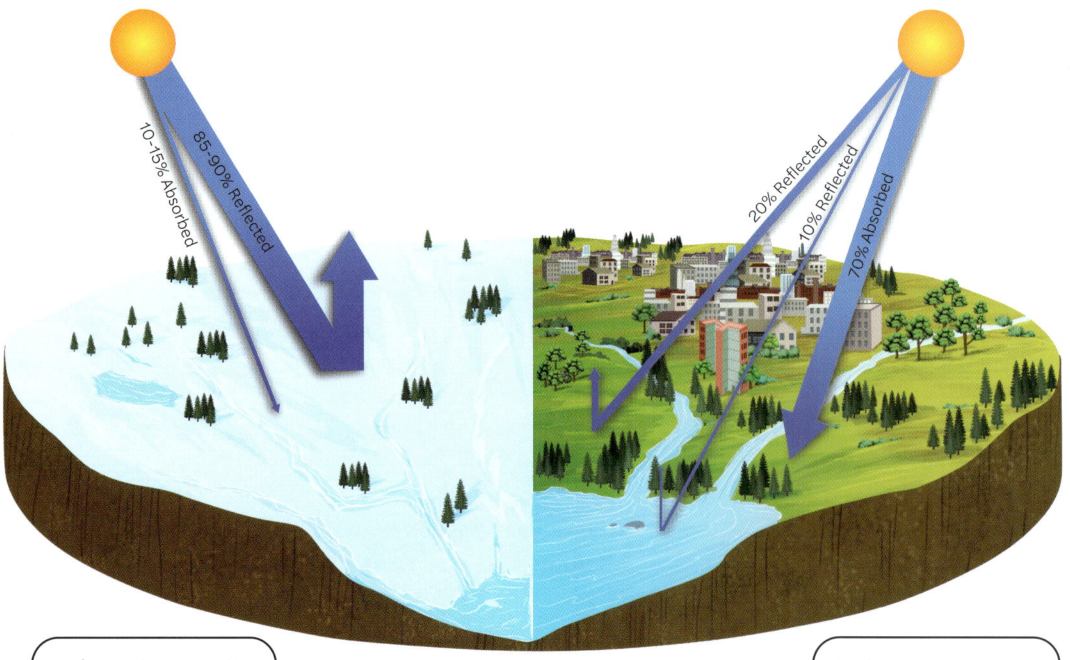

Source 1

Albedo refers to the amount of heat reflected or absorbed by the Earth's surface. White or light surfaces reflect more heat, generally resulting in cooler climates, while dark surfaces absorb more heat, resulting in warmer climates. Oceans are estimated to absorb up to 90 per cent of heat, meaning that ocean temperatures will warm, resulting in more ice loss and therefore warmer global temperatures.

Possible future sea levels for different greenhouse gas pathways

Change in sea level (metres): Extreme; High; Intermediate-high; Intermediate; Intermediate-low; Low; 2000 sea level

Tide-gauge observations; Satellite observations

Year: 1800, 1850, 1900, 1950, 2000, 2050, 2100

Source 2

Change in sea level over time (m) and expected changes based on predictions of greenhouse gas emissions in the future

Source: NOAA Climate.gov

Learning ladder G1.10

Show what you know

1 Describe the link between global warming and climate change.

2 Summarise why global warming is an issue when considering environmental *change* on any *scale*.

3 Suggest why the image of the stranded polar bear has been used to represent global warming.

Digital and geospatial technologies

Step 1: I can interpret different map types using cartographic conventions

4 Source 4: How does this map help geographers understand the *interconnection* between sea level rise and regional environmental *change*?

Step 2: I can construct paper maps using correct cartographic conventions

5 Research five *regions* that may undergo significant environmental *change* such as flooding or loss of habitat as a result of sea level rise and locate them using a blank world map.

Step 3: I can access and use geospatial technology platforms such as GIS

6 Access: http://mea.digital/GHV10_G1_3.

a Read the 'background' information. Describe the data that is available on this resource.

b Explore a few of the places highlighted on the home page. How are base maps and layers used to illustrate patterns?

Step 4: I can manipulate data using digital and geospatial technologies

7 Access: http://mea.digital/GHV10_G1_3 again. Select a location and then, using the 'manual' tool, alter the sea level rise. Comment on the different outcomes for *place* based on varying sea level heights. Suggest what could be done to prepare for the potential scenarios.

Source 3

The iconic image of a polar bear stranded on floating sea ice is often used to represent the impacts of global warming.

Sea level change (1993–2019)

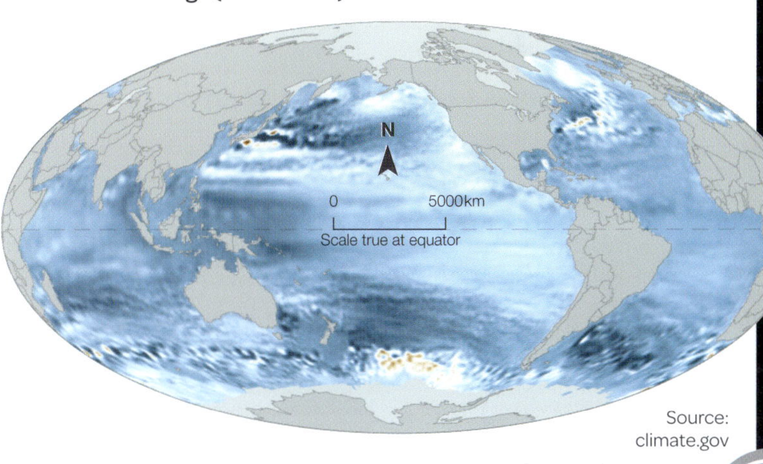

Source: climate.gov

Change in sea level (cm)

−20 0 20

Source 4

Global sea level change

HOW TO

SHEEPT, page 156
Answering questions, page 162

Why is environmental change a political issue?

The rate at which global temperatures are increasing is directly linked to the **greenhouse gas** emissions we produce. The changes we make to future climate will mean shifts in trade, agriculture and liveability of world regions, with potential to devastate national productivity and economies.

In 2018, the Intergovernmental Panel on Climate Change (IPCC) released 'The Special Report on Global Warming of 1.5°C'. The report cited over 6000 scientific references and concluded that limiting global temperature increases to 1.5°C would mean that people and places would have more time to adapt to current climatic changes and allow for a more sustainable future. In order to meet this target, we would need to reduce global net **emissions** of carbon dioxide by over 45 per cent by 2030 and reach 'net-zero' emissions globally by 2050.

This is not just a matter of reducing our output of carbon dioxide but also removing historical levels from the atmosphere. 'The decisions we make today are critical in ensuring a safe and sustainable world for everyone, both now and in the future', said Debra Roberts, Co-Chair of IPCC Working Group II. 'This report gives policymakers and practitioners the information they need to make decisions that tackle climate change while considering local context and people's needs. The next few years are probably the most important in our history', she said.

This report provides another example of the conflict of sustainability. Australia's number one export product is mineral fuels, including oil. Our exports are worth US$88.9 billion per annum, which equates to 32.6 per cent of total exports.

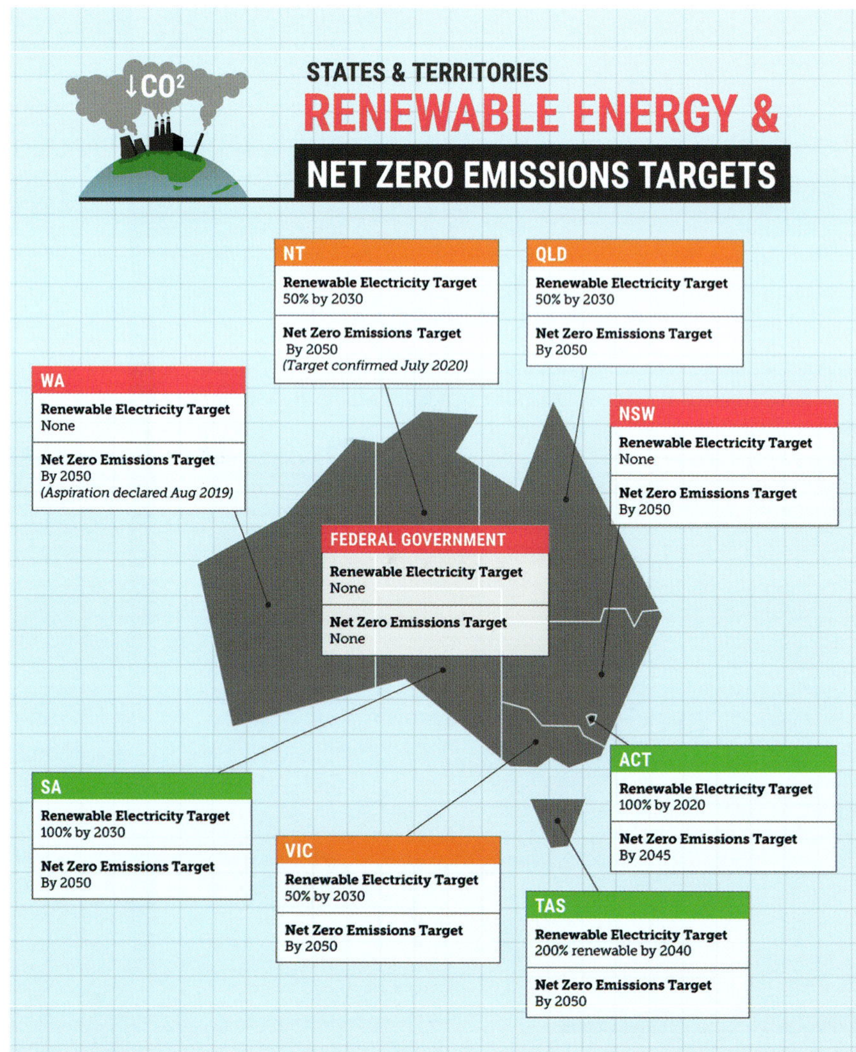

Source 1

Australia's emission targets by state

Source 3

A woman looks at the Ross River flooding over Aplins Weir in Townsville on 2 February 2019. Flood waters were released from the Ross River Dam to prevent the dam wall failing.

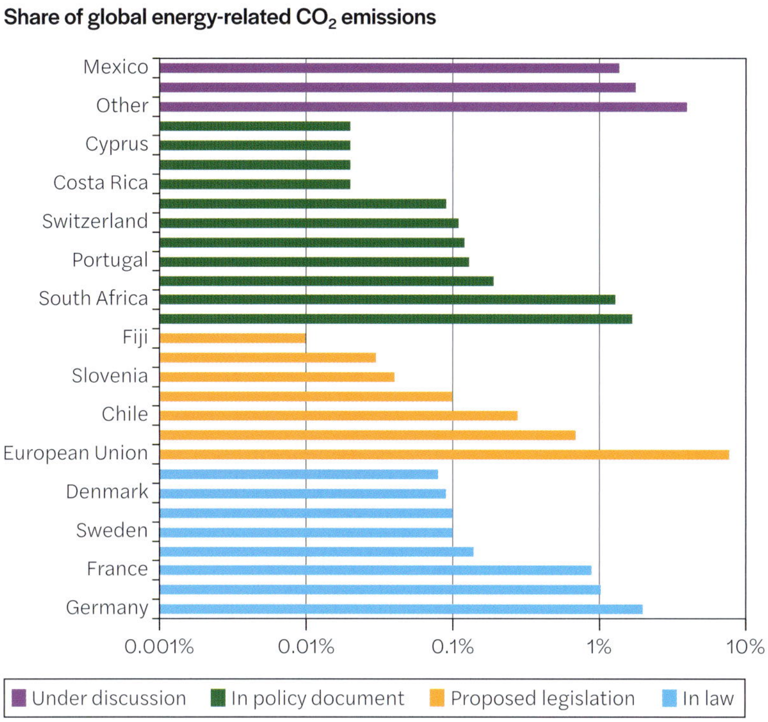

Our biggest buyers are China (32.7 per cent) and Japan (9 per cent). Mining also employs 2 per cent of Australia's total workforce and, since 2015, employment in this sector has grown by over 2.3 per cent. As a result, governments and policymakers need to find a balance between maintaining economic stability and ensuring we meet our contributions to reducing the impacts of global warming. This will become increasingly more important as our trade partners, such as Japan, begin to move away from using fossil fuels and work towards their own net-zero emissions goals.

Share of global energy-related CO_2 emissions

Mexico
Other
Cyprus
Costa Rica
Switzerland
Portugal
South Africa
Fiji
Slovenia
Chile
European Union
Denmark
Sweden
France
Germany

0.001% 0.01% 0.1% 1% 10%

Under discussion | In policy document | Proposed legislation | In law

Source: Internation Energy Agency

Source 2

Countries that have announced goals to meet net-zero carbon emissions by 2050.

Learning ladder G1.11

Civics and citizenship

Step 1: I can identify topics about society

1 Summarise the goals of 'net-zero' emissions.

Step 2: I can describe societal issues

2 Outline some conflicts that occur between reducing national carbon emissions and maintaining societal expectations.

Step 3: I can explain issues in society

3 Explain why Australia may be forced to explore more renewable trade options in the future.

Step 4: I can explain different points of view

4 Explain why some countries have committed to a net-zero impact by 2050, while others are hesitant. What is Australia's position?

Step 5: I can analyse issues in society

5 Write a letter to parliament outlining your position on Australia's current approach to managing carbon dioxide emissions and suggest some alternative solutions.

Answering questions, page 162

How do we reconcile environmental worldviews and management?

When European farmers first colonised Australia, they approached agriculture with the knowledge and traditions they used in their homelands. They dug irrigation channels and bred reliable species to produce goods for export and to support the growing colonies.

With population growth and the rainfall variations caused by the El Niño and La Niña events, problems soon became prevalent: crops failed, extended droughts caused hardships and soil compaction, salinity increased, while overcultivation and overgrazing saw the arable land become increasingly degraded.

The patterns of land degradation are widespread throughout areas where Europeans settled in the Asia-Pacific, Africa and the Americas, where land was seen as a resource to be used as a driver of social and economic growth. This differs significantly from the environmental worldviews of many other cultures, faiths and traditions, particularly those of indigenous peoples, many of whom see land through holistic and non-material lenses and see themselves as custodians, rather than conquerors. These two differing views are known as **anthropocentric** (human-centred) and **ecocentric** (environment-centred), yet balancing economic livelihoods with effective environmental management remains an ongoing challenge, particularly in developing economies.

'Many of our herbaceous plants began to disappear from the pasture land; the silk-grass began to show itself in the edge of the bush track, and in patches here and there on the hill. The patches have grown larger every year; herbaceous plants and grasses give way for the silk-grass and the little annuals, beneath which are annual peas, and die in our deep clay soil with a few hot days in spring, and nothing returns to supply their place until later in the winter following. The consequence is that the long deep-rooted grasses that held our strong clay hill together have died out; the ground is now exposed to the sun, and it has cracked in all directions, and the clay hills are slipping in all directions; also, the sides of precipitous creeks-long slips, taking trees and all with them. When I first came here, I knew of but two landslips, both of which I went to see; now there are hundreds found within the last three years.'

Source 1

Letter from John G. Robertson to Lt. Governor Charles La Trobe in 1853, referring to the Casterton and Coleraine areas of southwest Victoria

Source 2

The Menominee River in northeast US

Source 3

Members of the Menominee Nation at a traditional gathering

People's differing environmental worldviews produce different approaches to environmental management. Engagement with traditional owners is gaining increasing recognition in many places around the world, including forest management with the Menominee Nation in Wisconsin, USA, New Zealand's Whanganui River and Te Urewera National Park with the Māori and in South Australia, at the mouth of the Murray River, with the Ngarrindjeri Nation. Such partnerships help to strike a balance between economic benefits, cultural identity and the preservation and conservation of landscapes.

Learning ladder G1.12

Show what you know

1 Define 'anthropocentric' and 'ecocentric' worldviews.

2 Using examples, explain how the way we see the *environment* affects the ways we use it.

Changes and implications

Step 1: I can identify that changes occur in the characteristics of places over time

3 How does human usage *change* natural environments?

Step 2: I can describe how places have changed over time

4 Source 1: Describe how southwestern Victoria began to *change* soon after the arrival of Europeans.

Step 3: I can explain the causes behind the change over time in a place

5 Using a specific example, such as deforestation, create a flow chart to demonstrate how the natural *environment* has changed. Consider both the impacts, such as how it might affect flora, fauna and the physical *environment*, and the causes behind it.

Step 4: I can make predictions and outline consequences of change over time

6 Source 2: Suggest how engaging with the traditional owners of the Menominee River *region*, the Menominee Nation, could help provide a balanced approach to environmental management. What might be the likely outcomes for people and the *environment*?

HOW TO

Answering questions, page 162

How do we balance societal and environmental sustainability?

In 2019, as it has been since the establishment of the Australian colonies, natural resources comprised the bulk of Australia's exports, primarily raw minerals and agricultural products.

Our vast natural resources fuel key growth in economic development in many regions of the world, especially across the Asia-Pacific region. Combined with population growth – Australia's population has risen from around 5.4 million in 1920 to around 25.5 million in 2020 – urban expansion and the rise in commercial agriculture, as well as the economic value of our natural resources, have put increasing strain on sustainability.

Balancing the needs of the environment and the economy plays a key role in the changing nature of the Australian workforce. Industries that fuel the export of raw minerals and agricultural products continue to see job growth, but new career opportunities are expanding in areas such as renewable energy, green design and innovative agriculture, as well as in the marketing, ICT and education fields that support innovation in sustainability.

Innovative agriculture offers significant opportunities for career development that support both economic growth and environmental sustainability. The use of hydroponics, drip-feed irrigation, organic produce and new food markets are areas of focus.

Renewable energy involves the generation and distribution of clean energy, supplying low emissions power. In Australia, the introduction of solar panels and wind turbines, along with biomass and hydroelectric power, are attracting new investment.

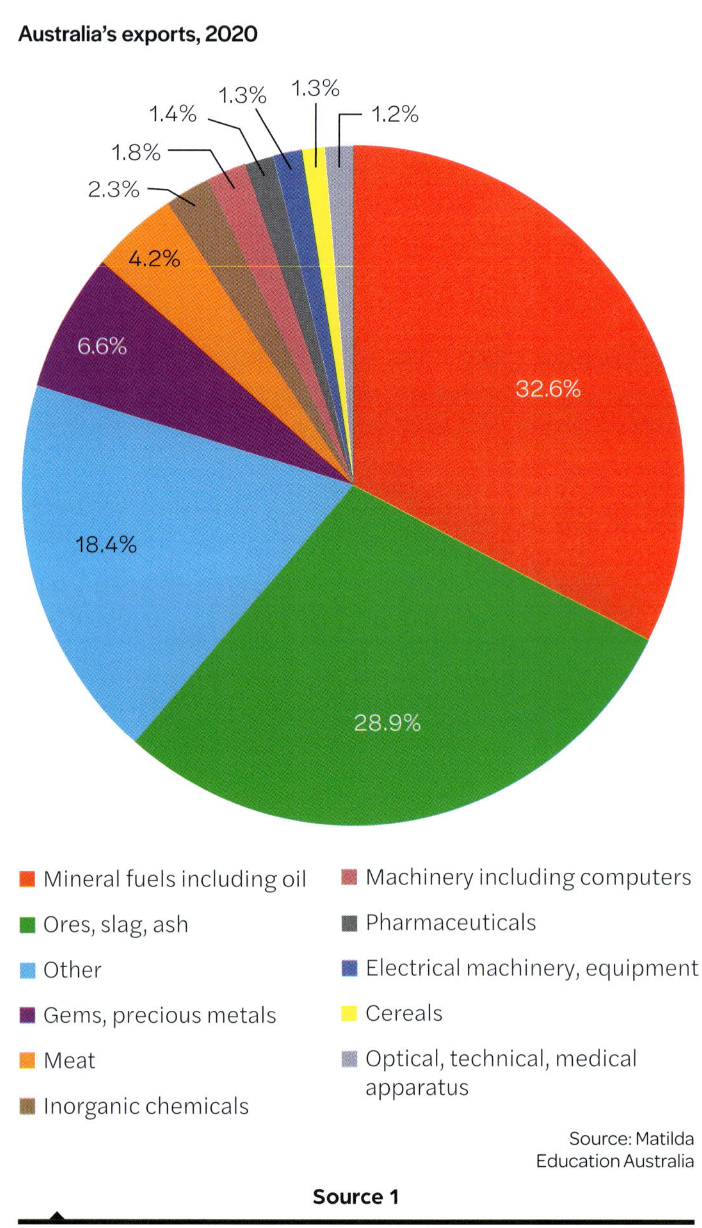

Source 1

Australia's top 10 exports in 2020 make up 81.6 per cent of all exports.

Source 2

New career opportunities are expanding in areas such as renewable energy.

Green design refers to the changing ways in which construction and design can incorporate renewable energy elements and innovative features to minimise or mitigate environmental impact. In urban areas, this can include the use of porous paving, water recycling or barriers to reduce noise pollution. Windcatchers – architectural features designed to channel natural winds for cooling and reduce the need for air-conditioning – are increasingly common, as are other design elements, such as rooftop gardens.

Source 3

The Parkroyal tower in Singapore features 12-storeys of solar-powered sky gardens. The building uses automatic light, rain and motion sensors, rainwater harvesting and recycling mechanisms.

Learning ladder G1.13

Show what you know

1 How has Australia's population *changed* over the past century?

2 Why are Australia's primary exports potentially unsustainable in the long term?

Economics and business

Step 1: I can recognise economic information

3 Source 1: Identify the percentage of Australia's top 10 exports in 2020 that are based on natural resources.

Step 2: I can describe economic issues

4 Which new career areas are developing as the push to balance economic and environmental *sustainability* increases?

Step 3: I can explain issues in economics

5 Australia's commitment to reduce its carbon emissions will impact the future jobs market. Explain which fields are likely to be, directly and indirectly, negatively impacted.

Step 4: I can integrate different economic topics

6 Create three mind maps, one each for innovative agriculture, renewable energy and green design. Record as many different jobs that might be linked to each, directly or indirectly. For example, solar panels need people to design, build, transport, market, sell, install and maintain them.

Step 5: I can evaluate alternatives

7 'The end of the mining boom means the end of the Australian economy.' To what extent do you agree with this statement? Refer to alternative energy and new career opportunities in your response.

HOW TO Answering questions, page 162

Place study
G2.5

G1.14

Why is environmental management important?

As the human population continues to grow and our need for resources increases, we need to plan and prepare for a more sustainable future.

Improvements in technology have allowed us to explore renewable energies, new trade opportunities and more sustainable farming practices. **More economically developed countries (MEDCs)** are, on average, the most energy hungry **per capita**. For example, in 2019 US citizens used an average of over 79 897 kilowatts of energy per person, while those from Niger only used 457 kilowatts per person. However, in 2015 the USA only invested 0.2 per cent of its GDP towards renewable energy alternatives. In order to make global change, we need to shift our use of environmental space and resources on local and regional scales.

Managing environmental change does not mean we have to significantly alter our lives. In many cases we are able to adapt products and waste or change land use to both benefit communities and manage environmental sustainability.

Source 1

The municipal corporation in Chennai, India is constructing infrastructure to increase the number of pedestrians and cyclists by 40 per cent. This will help reduce motor transport and its pollution. The city has also partnered with SmartBike to provide electric bicycles and shaft transmission (no chain) NextGen bicycles.

Source 2

Google Street View cars have been measuring and mapping air quality since 2015, allowing governments, scientists and policymakers to monitor smog, nitrogen oxides, carbon dioxide and methane in the atmosphere and plan or respond in targeted ways.

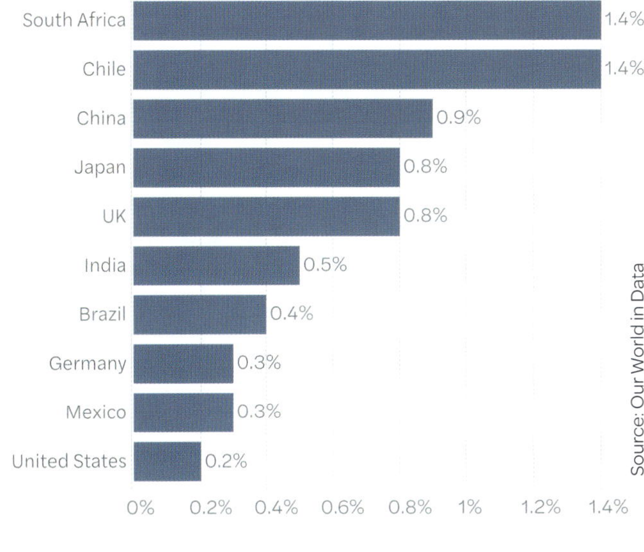

Source 3

The percentage of a country's GDP spent on renewable energy in 2015

Learning ladder G1.14

Show what you know

1 Summarise why environmental management is important for environmental *sustainability*.

2 As a class, brainstorm some ways to manage human environmental impact.

3 Go out of the classroom and explore the school grounds. Create a list or a sketch outlining how your school could operate more sustainably.

Communicate data

Step 1: I can list primary and secondary methods useful for my study

4 Source 1: Suggest some primary data that researchers could collect in Chennai, India to assess the success of their strategy to reduce carbon emissions.

Step 2: I can successfully use data collection methods

5 Source 3: Given many countries are committed to net-zero emissions by 2050, predict how renewable energy investment may change in the future.

Step 3: I can filter collected data

6 Access: http://mea.digital/GHV10_G1_4. Choose a category of interest (building, fashion, transport etc.) and explore some of the sustainable solutions around the world. Select one project and summarise how it aims to manage environmental change.

Step 4: I can organise data collected according to relevance for a research question

7 Access: http://mea.digital/GHV10_G1_4 again. You are selected as a geographic consultant to help develop technology to reduce carbon emissions and environmental impact. In a small group, brainstorm some environmental issues in your *region*. Using the Sustainia 100 projects as inspiration, invent a potential solution to this issue and present your idea as if it was another page in this report.

SHEEPT, page 156
Sketches and annotating, page 158
Answering questions, page 162

How do First Nations peoples manage environmental change?

For over 65 000 years, First Nations peoples managed the land using a range of practices including fire. Known as **cultural burning**, this technique is conducted in particular seasons and uses a **mosaic** of small-scale, 'cool' fires or controlled, low intensity burns to manage human risk and natural fuel build up, while ensuring there is minimal impact to forest density, resources or animal life.

Such land management practices are strongly interconnected with First Nation peoples' sense of place and traditional understanding of the positive relationships between fire, land productivity, regeneration and the needs of native wildlife. Cultural burning, in a way, aims to enhance and heal the identity of a place, rather than alter it for human needs and protection. In many cases, cultural burning is not primarily conducted as a fuel reduction method, but instead as a way of caring for the landscape.

Currently, back burning or controlled burns tend to be more intensive than cultural burns, and, while they act to clear fuel, they also may impact tree canopies, increasing the rate of soil moisture loss and invasive plant and weed species growth in the cleared space. These techniques have also been largely criticised, especially after the devastating 2019–2020 fires where over 10 million hectares in southern Australia were burnt. In 2020, the Royal Commission into National Natural Disaster Arrangements, often called the 'bushfire royal commission', identified that First Nations land management is vital to reducing the risk of disastrous bushfires. In fact, scientists have found that cultural burning practices conducted during the dry-season in northern Australia has decreased the extent of fires by around 30 per cent compared to previous years when cultural burning was not conducted.

Discussions at the 2020 Royal Commission revealed, however, that in the past First Nations peoples' voices and knowledge had been ignored when dealing with land management and burning practices. During the commission, Mr Eckford-Williamson (an Euahlayi man and ANU researcher) stated that First Nations peoples must take part in any cultural land management practices and that 'the thing that makes it cultural burning is controlled by Aboriginal people'. 'Cultural land management is not an add-on or an enhancement,' he said. 'It's not a practice that can simply be grafted onto the regime of non-Indigenous land managers.'

In 2019, the first cultural burn in over 170 years was conducted north of Bendigo. Local First Nations groups have found that, since then, cultural burning has enhanced local sites, allowing for the regeneration of key native species such as Kangaroo grass, which was used to make some of the world's first bread products. As our climate warms because of global warming, cultural fires may provide a safer and more productive way of managing our land to reduce bushfire risk and ensure the safety of modern populations and infrastructure.

Source 1

This image shows prescribed grass burns at Fish River, NT, as part of the carbon management plan. The Nature Conservancy is working with First Nations rangers to manage the land through traditional burning regimes.

Learning ladder G1.15

Show what you know

1 Define 'cultural burning' and describe the process of this traditional land management technique.

2 'First Nations voices are vital to the success of land management in Australia.' Discuss this statement as a class.

Patterns and interconnections

Step 1: I can provide short explanations for patterns and interconnections

3 Describe the key differences between cultural burning and non-Indigenous land management techniques.

Step 2: I can explain patterns and interconnections

4 Suggest how cultural burning and other First Nations land management practices may help reduce bushfire risk for communities in future generations.

Step 3: I can use data to support explanations of patterns and interconnections

5 Discuss the following statement in groups and research data to support your opinions. 'Using traditional practices will better equip us for managing landscapes in the future.'

Step 4: I can use relevant sources to research further reasons for patterns and interconnections

6 Research the disastrous fire event that occurred in 2019–2020. Why was there a Royal Commission in 2020 and what were its major findings?

Answering questions, page 162

Masterclass

Learning ladder

Work at the level that is right for you or level-up for a learning challenge!

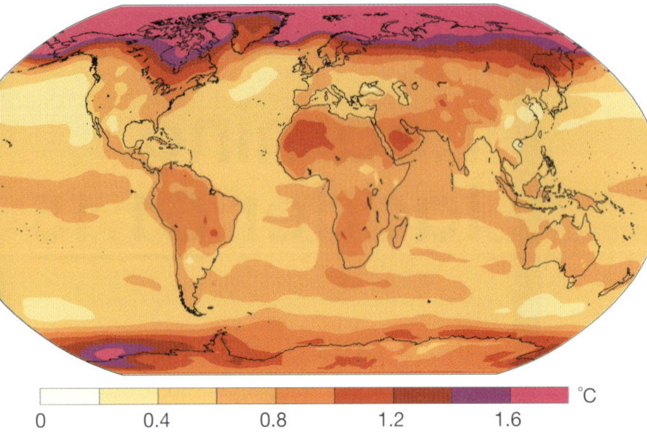

Source 1

Change in yearly temperature as a result of human activity during the baseline period 1986–2005

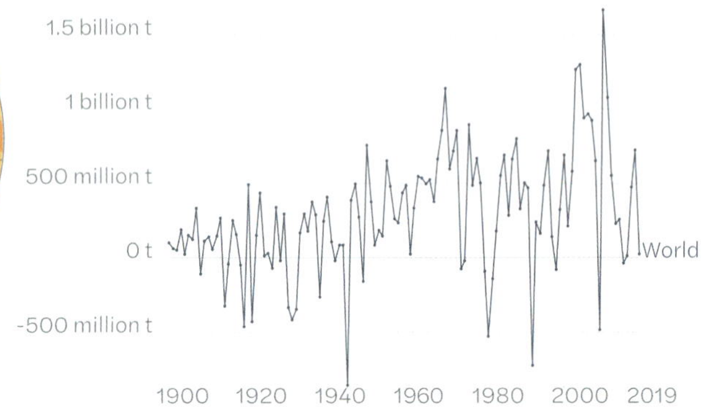

Source 2

Annual change in carbon dioxide emissions and atmospheric levels from 1900–2019

Source 3

Environmental change occurring in an urban region

Step 1

a I can identify spatial distributions and patterns

Source 1: Identify two countries or *regions* that are predicted to have the highest temperature increase over time.

b I can provide short explanations for patterns and interconnections

Explain why carbon emissions made locally have global impacts.

c I can identify that changes occur in the characteristics of places over time

Identify three environmental changes that have occurred in Australia over time.

d I can list primary and secondary methods useful for my study

Source 3: Create a list of reliable secondary sources to collect data on the impact of rubbish on natural environments.

e I can interpret different map types using cartographic conventions

Suggest whether a choropleth map or cartogram would be more useful for illustrating global sea level rise.

Step 2

a I can use data to quantify spatial distributions and patterns

Source 2: Quantify the following statement: 'Carbon emissions have increased over time'.

b I can explain patterns and interconnections

Source 2: Explain why atmospheric carbon has increased over time.

c I can describe how places have changed over time

Using the correct conventions, complete a field sketch with clear annotations highlighting changes that have occurred in your local *region* over time.

d I can successfully use data collection methods

Source 1: List two methods that may have been used to collect data to create this map.

e I can construct paper maps using correct cartographic conventions

Source 1: Using tracing paper, complete BOLTSS on the map.

Step 3

a I can describe spatial distributions and patterns

Describe the *spatial association* between temperature increases and the occurrence of drought in Australia.

b I can use data to support explanations of patterns and interconnections

Source 2, page 33: Quantify the predictions of global sea level rise for low, intermediate and extreme scenarios.

c I can explain the causes behind the change over time in a place

Source 1: Explain why the poles may experience the greatest temperature changes over time.

d I can filter collected data

Source 3: List evidence showing environmental changes that have occurred over time.

e I can access and use geospatial technology platforms such as GIS

Access http://mea.digital/GHV10_G1_5 and log in as a 'guest'. List the data that is able to be overlaid to identify changes in climate on this map. Outline why this data may be helpful in studying environmental *change*.

Masterclass

Step 4

a I can use data to support exceptions to spatial distributions and patterns

Source 3, page 21: Atmospheric carbon has naturally fluctuated over time. Use quantitative data to show how human activity has altered this natural pattern.

b I can use relevant sources to research further reasons for patterns and interconnections

Using reliable sources, investigate traditional land management practices from another country or *region*. Summarise how *change* in management over time has led to positive or negative outcomes for the *environment*.

c I can make predictions and outline consequences of change over time

Source 2: Predict the future patterns of both carbon emissions and atmospheric carbon.

d I can organise data collected according to relevance for a research question

As a class, undertake fieldwork to investigate the following question: 'How is Australia's *environment* likely to *change* in the future?'. List the data presented in G1 that may help answer this research question.

e I can manipulate data using digital and geospatial technologies

Access http://mea.digital/GHV10_G1_6 and log in as a 'guest'. Turn on two different layers and change the opacity so you can easily see both. Comment on any interactions or patterns that occur in one world *region*.

Step 5

a I can identify multiple spatial distributions and patterns

Source 3: Human activity has significant impacts on a *place*. Identify any human environmental interactions that are evident in this image.

b I can interpret causes of patterns and interconnections

Summarise the *interconnection* between carbon emissions from human activities and global warming.

c I can interpret data to quantify predictions based on research

Research a current government policy or law that aims to reduce the impacts of deforestation in Australia. Evaluate the likely effectiveness of this strategy, using data to quantify your response.

d I can evaluate the success of research methods

Predict how geospatial technology will be helpful in monitoring climatic and environmental *change* in the future.

e I can draw conclusions from geographical information in digital and geospatial technologies

Access: http://mea.digital/GHV10_G1_7. Explore the data available and write a short concluding paragraph that summarises the ideas, concepts, key terms and data discussed through G1.

Capstone

How can I understand changes and challenges?

In this chapter, you have learned a lot about changes and challenges. Now you can put your new knowledge and understanding together for the capstone project to show what you know and what you think.

In the world of building, a capstone is an element that tops off a building or wall. That is what the capstone project will offer you, too: a chance to top off and bring together your learning in interesting, critical and creative ways. You can complete this project yourself, or your teacher can make it a class task or a homework task.

mea.digital/GHV10_G1

Scan this QR code to find the capstone project online.

Place studies of change and management G2

WILL FIJI DISAPPEAR AS SEA LEVELS RISE?

page 54

digital and geospatial technologies

page 58

HOW DO EXTREME WEATHER EVENTS CAUSE ENVIRONMENTAL CHANGE?

changes and implications

page 66

HOW DOES HUMAN MANAGEMENT IMPACT PLACE?

getting involved

page 72

HOW CAN I MAKE A DIFFERENCE?

How can I learn about change and management using place studies?

It is difficult to understand the impact of environmental change without investigating places. In Geography, we use case studies of particular places to deepen our knowledge. Environmental change is a result of seasonal and long-term climate activity, interactions with animal and plant species, and human activity. The impacts we have on a local scale can have global consequences.

learning ladder

Step	Spatial distributions and patterns	Patterns and interconnections	Changes and implications
step 5	**I can identify multiple spatial distributions and patterns** I can take my PQE one step further to find links or relationships that exist in place studies of change and management.	**I can interpret causes of patterns and interconnections** I can use multiple sources to find links or relationships that exist in place studies of change and management and can explain 'Why?'.	**I can interpret data to quantify predictions based on research** I can use external data from research as evidence of the positive and negative impacts of a change I have predicted.
step 4	**I can use data to support exceptions to spatial distributions and patterns** I can use data to answer 'Why?' about the exceptions identified in a PQE analysis of place studies.	**I can use relevant sources to research further reasons for patterns and interconnections** I can use sources other than this textbook to further research patterns I observe in place studies.	**I can make predictions and outline consequences of change over time** I can use my knowledge of natural processes and world regions to make an educated guess about the positive and negative impacts of change in place studies.
step 3	**I can describe spatial distributions and patterns** I can describe patterns, quantify them and point out exceptions (PQE) to describe place studies.	**I can use data to support explanations of patterns and interconnections** I can use data from a map or graph to explain patterns I observe in place studies.	**I can explain the causes behind the change over time in a place** I can use my knowledge of natural processes and world regions to explain why changes may occur over time in place studies.
step 2	**I can use data to quantify spatial distributions and patterns** I can read data and use it to measure key trends on a map or graph about place studies.	**I can explain patterns and interconnections** I can identify social, historical, economic, environmental, political and technological (SHEEPT) factors to help me explain place studies.	**I can describe how places have changed over time** I can use specific examples to describe changes over time in place studies.
step 1	**I can identify spatial distributions and patterns** I can find key trends on a map or graph about place studies.	**I can provide short explanations for patterns and interconnections** I can write descriptions of patterns and interconnections that I find in place studies.	**I can identify that changes occur in the characteristics of places over time** I can read information and answer questions about changes over time in place studies.

Source 1

Seasonal variation, long-term changes to climate and human activities all lead to environmental change.

I can evaluate the success of research methods
I can look back and comment on the data collection methods I used, and evaluate how successful they were in helping me answer a research question about place studies.

I can organise data collected according to relevance for a research question
I can review the data I have collected in the field and display it using graphs, tables, annotations and captions.

I can filter collected data
I can review my collected data and select the most relevant data to answer a research question about place studies.

I can successfully use data collection methods
I can use primary and secondary data collection methods in the field and classroom to investigate place studies.

I can list primary and secondary methods useful for my study
I can create a checklist of methods to investigate place studies and categorise them as primary or secondary methods.

Communicate data

I can draw conclusions from geographical information in digital and geospatial technologies
I can interpret and analyse patterns by using different layers and features on geospatial technology platforms.

I can manipulate data using digital and geospatial technologies
I can work with layers and other features on geospatial technology platforms to further explore data and interconnections.

I can access and use geospatial technology platforms such as GIS
I can use geospatial technology platforms to explore data and find patterns.

I can construct paper maps using correct cartographic conventions
I can use a pencil, paper and ruler to construct a map that follows BOLTSS conventions.

I can interpret different map types using cartographic conventions
I understand data found in different types of maps and graphs and use the data to answer questions about place studies.

Digital and geospatial technologies

Spatial distributions and patterns

1 Predict which *regions* of the world are most vulnerable to sea level rise as a result of glacial melt.

Patterns and interconnections

2 Outline what impacts the environmental *change* seen in Source 1 may have on people and *place*.

Changes and implications

3 Suggest one positive and one negative impact human management might have on *environments*.

Communicate data

4 Case studies involve exploring and analysing real-world geographical issues. Suggest two reliable data sources that may help you quantify environmental *change* and management in different world *regions*.

Digital and geospatial technologies

5 Outline how geospatial technology would be useful in predicting, planning for and managing droughts in Australia. Research an example of a real-life application.

How is urbanisation in Lagos having environmental impacts?

Lagos is a low-lying, coastal state of Nigeria, Africa. While it only accounts for 0.4 per cent of Nigeria's total land area, it is one of the fastest growing places on Earth. In 1950, Lagos was home to just over 300 000 people, but in 2020 its population was over 14 million people. Rapid urbanisation of the region has put huge pressure on the environment.

Lagos is characterised by its lagoons, swamps and floodplains. More than one-fifth of people living in Lagos are at risk of or are living in **poverty** and so are forced into high-density slums built on stilts to avoid flood waters. Currently, Lagos produces over 13 000 metric tonnes of waste per day. Mismanagement of waste means people in slums are exposed to diseases such as Lassa fever.

The volume of waste also has a major effect on the natural environment. Research shows that waste pollutants affect the ability for animals to grow and reproduce and humans may be at risk of consuming harmful chemicals by eating fish from local waters.

Rapid urbanisation also puts pressure on environmental resources. It is estimated that Lagos requires over 2044 megalitres of water per day, but only 794 megalitres of clean water can be produced. Much of the water collected is from groundwater contaminated by industry and high-density living. Researchers predict that more than $2 billion would be needed to suitably upgrade water supply networks in Lagos to meet demand.

Source 1

Lagos, Nigeria

Nigeria

NIGER
CHAD
NIAMEY
Lake Chad
N'DJAMENA
Sokoto
Hadejia
Gongola
Kainji Reservoir
BENIN
NIGERIA
ABUJA
Niger
TOGO
LOMÉ
PORTO-NOVO
Lagos
CAMEROON
YAOUNDÉ
ATLANTIC OCEAN
Niger Delta
MALABO
Isla de Bioco
EQUATORIAL GUINEA

Source: Alamy.com

Source 2

A political map of Nigeria

Lagos generates over 40 per cent of Nigeria's GDP and has one of the highest densities of industry in the country. But while transportation is vital for trade, it is estimated that over 2 million cars travel around Lagos daily, with an average commute taking around 4 hours. This means a large amount of atmospheric particulate matter (PM) is released. The safe annual mean PM concentration level recommended by the World Health Organization (WHO) is 10 $\mu g/m^3$; however, Lagos has recorded levels of over 68 $\mu g/m^3$ – almost seven times the safe amount. These pollutants have been linked to more than 11 200 premature deaths and are also increasing the rate of global warming, with predictions suggesting that Nigeria will emit over 900 million tonnes of carbon per year by 2030 if it continues on its current **trajectory**.

Source 3

Lagos produces more than 13 000 metric tonnes of waste per day, the most of any Nigerian city.

Source 4

Lagos growth map

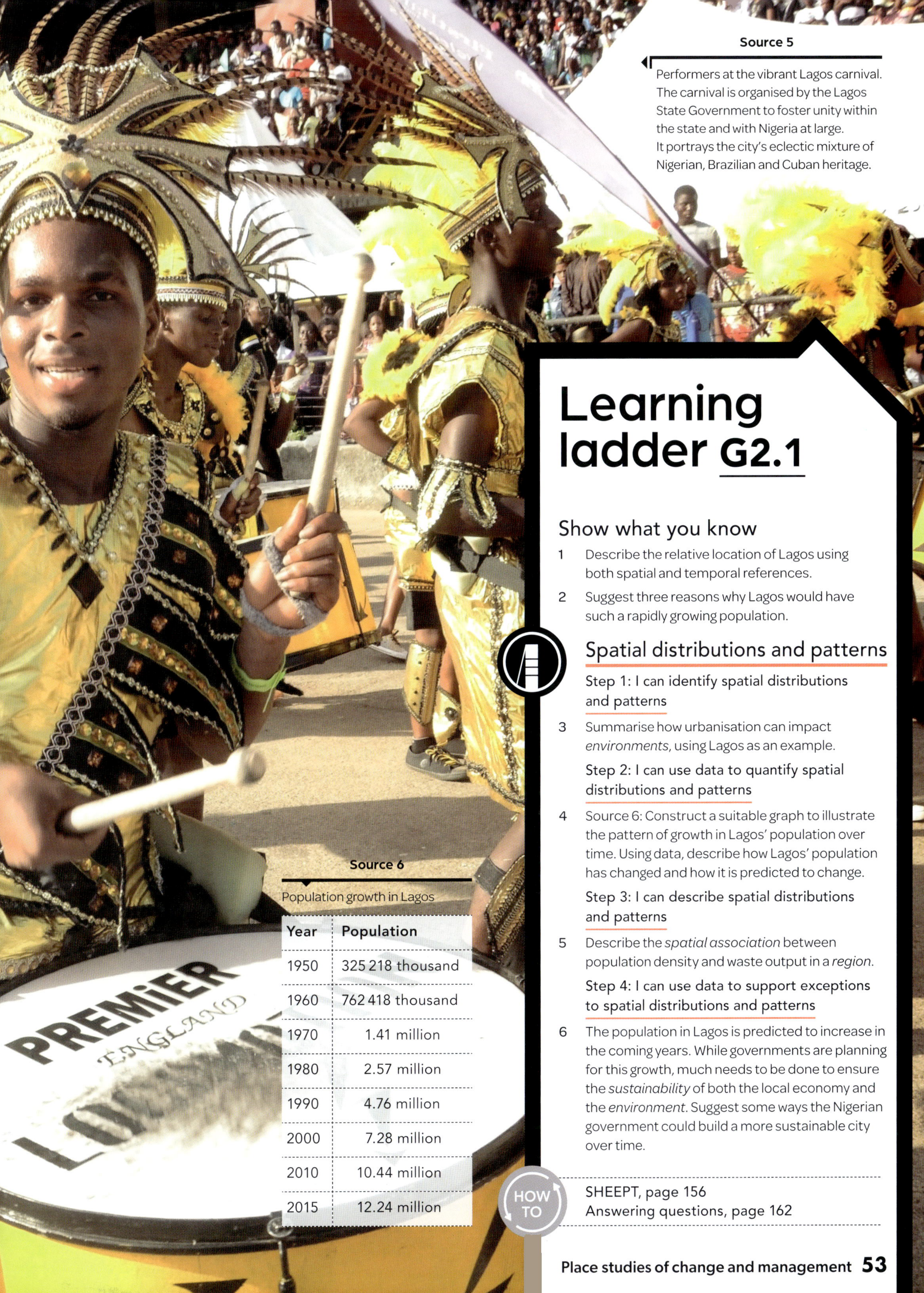

Source 5

Performers at the vibrant Lagos carnival. The carnival is organised by the Lagos State Government to foster unity within the state and with Nigeria at large. It portrays the city's eclectic mixture of Nigerian, Brazilian and Cuban heritage.

Source 6

Population growth in Lagos

Year	Population
1950	325 218 thousand
1960	762 418 thousand
1970	1.41 million
1980	2.57 million
1990	4.76 million
2000	7.28 million
2010	10.44 million
2015	12.24 million

Learning ladder G2.1

Show what you know

1 Describe the relative location of Lagos using both spatial and temporal references.

2 Suggest three reasons why Lagos would have such a rapidly growing population.

Spatial distributions and patterns

Step 1: I can identify spatial distributions and patterns

3 Summarise how urbanisation can impact *environments*, using Lagos as an example.

Step 2: I can use data to quantify spatial distributions and patterns

4 Source 6: Construct a suitable graph to illustrate the pattern of growth in Lagos' population over time. Using data, describe how Lagos' population has changed and how it is predicted to change.

Step 3: I can describe spatial distributions and patterns

5 Describe the *spatial association* between population density and waste output in a *region*.

Step 4: I can use data to support exceptions to spatial distributions and patterns

6 The population in Lagos is predicted to increase in the coming years. While governments are planning for this growth, much needs to be done to ensure the *sustainability* of both the local economy and the *environment*. Suggest some ways the Nigerian government could build a more sustainable city over time.

HOW TO

SHEEPT, page 156
Answering questions, page 162

Will Fiji disappear as sea levels rise?

Depending on how we manage our emissions, researchers estimate that up to 1 billion people could be **displaced** as a result of global warming by 2050. One of the main concerns in the South Pacific is sea level rise, which threatens to 'sink' low-lying islands.

The country of Fiji comprises 332 islands and in 2020 it was home to more than 896 445 people. Fiji is characterised by its 1129 kilometres of coastline, coral reefs and small traditional villages. Fiji is particularly susceptible to sea level rise because of its connection with and reliance on coastal regions for traditions, housing and income. The NOAA recently reported that the global sea level has risen by 16–21 cm between 1900 and 2016 and research has estimated that 4.5 per cent of Fiji's buildings will be flooded by a further 22 cm sea level rise.

Despite governments trying a range of measures such as building **sea walls**, years of flooding, coastal erosion and dangerous storms meant that, in February 2014, the village of Vunidogoloa moved 2 km inland. While Vunidogoloa was the first to relocate, another three villages have since moved as a result of the changes caused by global warming.

Source 1

Fiji is located in the South Pacific and consists of 332 islands

Fiji

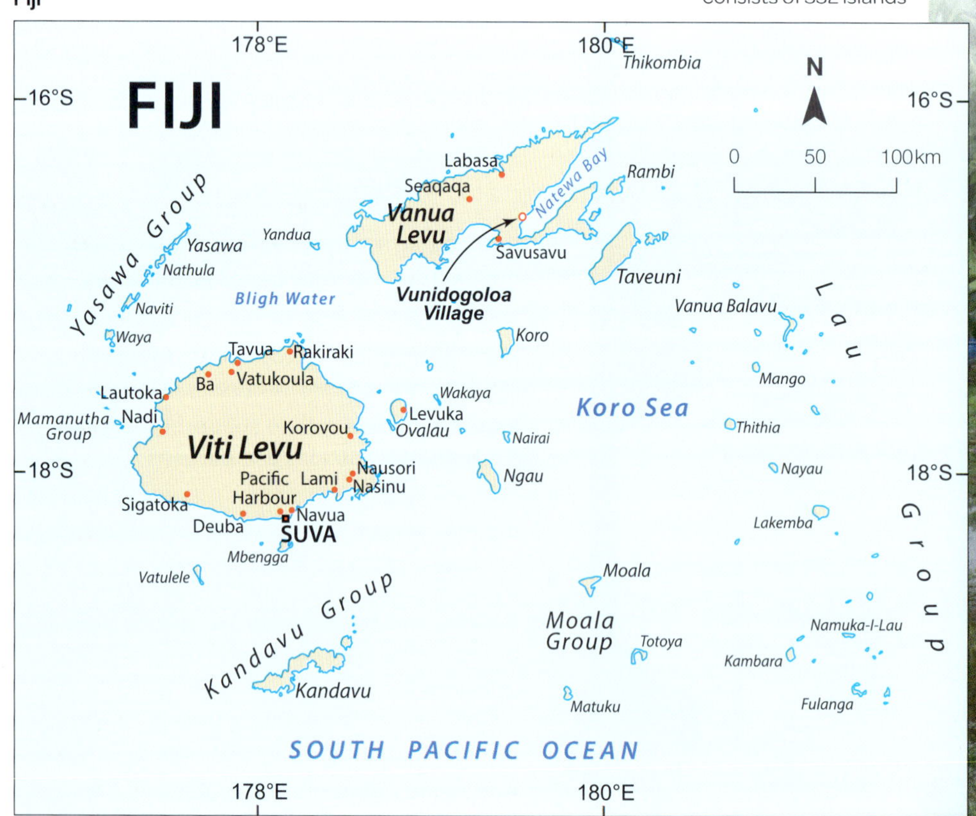

Source: Alamy.com/Custom Mapping Services

While relocation meant that villagers gained access to solar power, rainwater tanks and better facilities, residents complained that because houses were built quickly, they leak in the high-rainfall months. Studies have found that the move has also had a significant impact on the community's **sense of place**. People have been disconnected from their traditional and sacred lands and have restricted access to fishing sites, one of the dominant local food sources. Further, it is estimated that Fiji will lose up to 5 per cent of its GDP each year as a result of the changing environment, with extreme weather events such as tropical cyclone Winston in 2016 only making things worse. After the relocation of Vunidogoloa, the Fijian government established relocation guidelines and identified over 830 communities that will be vulnerable to the impacts of global warming, with 45 of these being earmarked for future relocation.

Since the Industrial Revolution, global temperatures have increased by at least 1°C. Global warming is not something that will occur in the future; it is a process that is causing environmental change now. Fiji only emits around 1 per cent of the world's carbon output, yet it currently faces some of the largest consequences of climate change. In order to combat enhanced climate change and reduce local impacts, we need to change global behaviours.

Source 2

Front cover of a magazine warning of the dangers to environments as a result of sea level rise

Source 3

Sea level is predicted to rise by over 30 cm by 2050, putting many cities and coastal low-lying regions at risk of flooding and inundation.

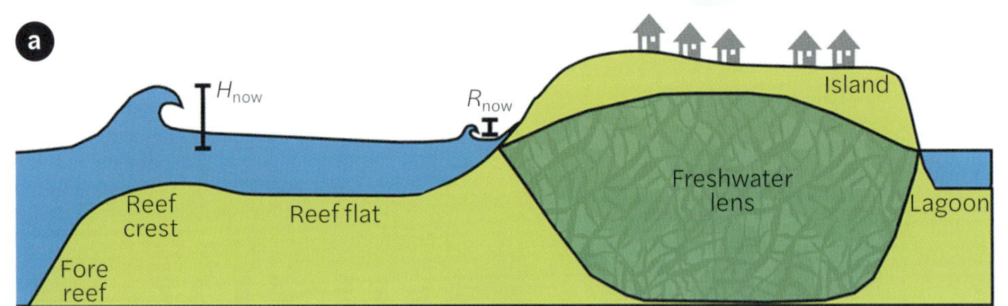

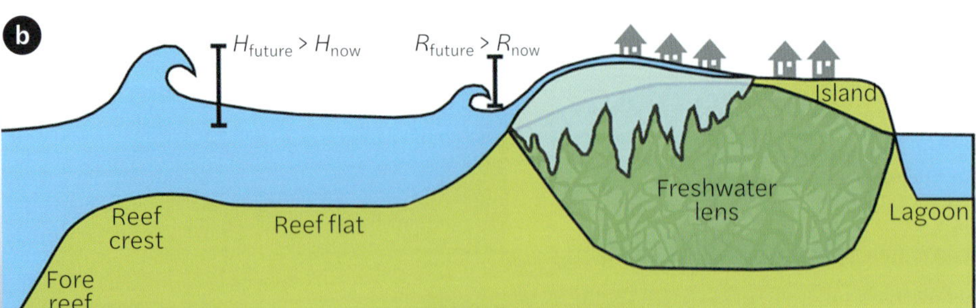

Atoll
Present sea level
Future elevated sea level
Present groundwater
Future groundwater
Salinised groundwater due to overwash

Source 4

Sea level rise is predicted to have larger impacts than just coastal town flooding. It impacts local freshwater sources, reduces land size for residents and affects coral reef systems

Source 5

The village of Vunidogoloa moved 2 km inland as a result of its vulnerability to rising sea levels.

Predicted sea level rise by 2030 in the Pacific region.

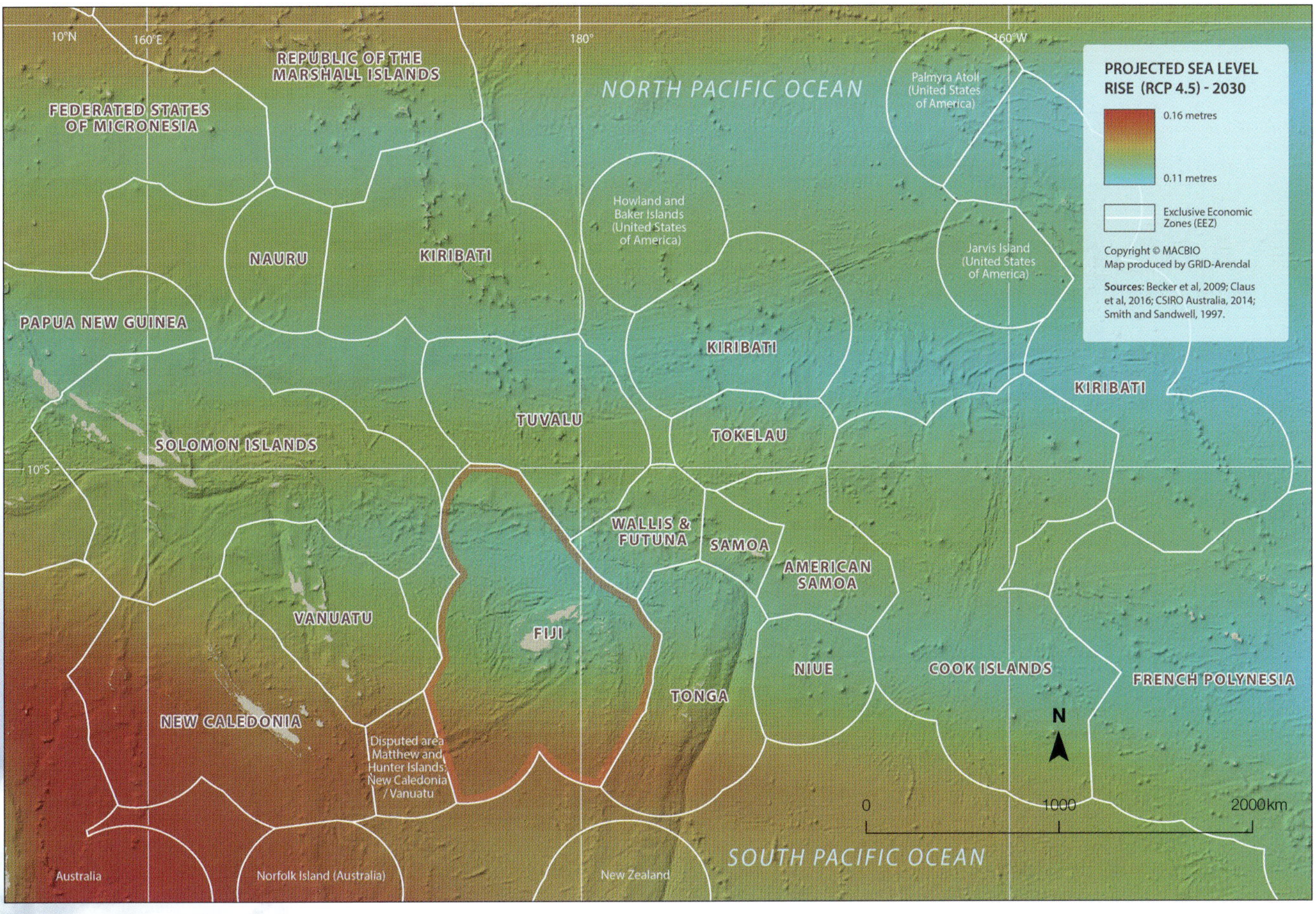

Source: Levi Westerveld

Source 6

This 2019 map shows projected sea level rise by 2030 in the Pacific region. Countries' Exclusive Economic Zones (EEZ) are shown in white. Fiji's EEZ is shown in red.

Learning ladder G2.2

Show what you know

1 a Fiji is a nation of 332 islands. Describe the **relative location** of Fiji.

b Source 1: State the approximate **absolute location** of Vunidogoloa.

2 Outline what is meant by the term 'sinking' when referring to islands and sea level rise.

Changes and implications

Step 1: I can identify that changes occur in the characteristics of places over time

3 Summarise how sea level rise causes environmental *change* in places such as Fiji.

4 Relocation was a last resort for the village of Vunidogoloa. Identify key characteristics that the relocation site would need to ensure the safety of the villagers.

Step 2: I can describe how places have changed over time

5 Source 6: Describe the *change* over time in Fiji's sea levels by 2030.

Step 3: I can explain the causes behind the change over time in a place

6 Source 1: Suggest why sea level rise may not impact all land masses equally. For example, why is Fiji more susceptible to the impacts of sea level rise than the USA?

7 Evaluate the effectiveness of Fiji's response to rising sea levels in Vunidogoloa.

Step 4: I can make predictions and outline consequences of change over time

8 Using data or examples, predict the future of island nations such as Fiji. Create a mind map to illustrate the likely outcomes for global trade, tourism and **displacement** or movement of people.

SHEEPT, page 156
Photo essays, page 160
Answering questions, page 162

How do extreme weather events cause environmental change?

Experts predict that global warming will create a dramatic increase in the frequency and intensity of future natural disasters. In Australia, we have already felt some of these impacts through decades of drought, out-of-season flooding and severe bushfires. These disasters bring about huge changes to the natural environment. While change is not always negative, such as the **regeneration** and **germination** of Australian tree species after fire, ongoing and unpredictable climate fluctuations may mean that environments are not able to adapt over time, resulting in mass **extinctions**.

Fires in California

Wildfires have always been a part of California's natural landscape. However, increased frequency of drought, heatwaves during spring and summer and reduced winter snowfalls means that the region is hotter, drier and more vulnerable to extreme fires. Global warming has led to longer and more unpredictable fire seasons, with more severe out-of-control fires three times more likely in future years than in 1970.

In August 2020, the fires in California, which had been burning for more than 50 days, earned a new classification as a **gigafire**, meaning it had spread over a million acres (404 700 hectares). In 2020 alone, California recorded five out of six of the largest fires in its history. More than 20 000 firefighters from all around the world were sent to fight the extreme fires and, unfortunately, the fires caused at least 31 deaths.

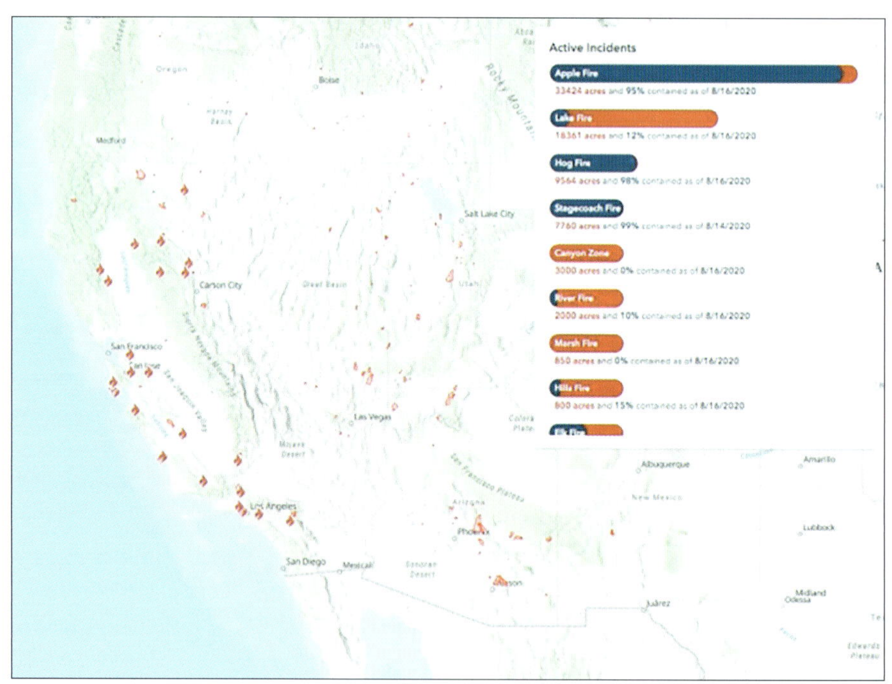

Source: Newsweek

Source 1

Screenshot highlighting the use of GIS for providing community members, decision makers and emergency services teams with up-to-date information about fire progress and predictions

Much of the environmental impact is not yet known. In 2021, teams of scientists and veterinarians were still rescuing and treating animals from burn sites. The wildfires may have significantly affected the native endangered pygmy rabbit, of which around only 50 individuals remain. The small rabbit's habitat was largely destroyed, along with nesting grounds for bird species such as the sage and sharp-tailed grouse. Plant populations such as the Coulter pine, which is **endemic** to the California region, are also under threat because of the fires.

Annual burned area has increased fivefold in California

Source 2

Change over time in fire activity in California (1972–2017). Fire intensity and damage levels have increased over time. Fire seasons have also extended from May to December.

Source 3

The California wildfires

Drought in the Horn of Africa

As a result of global warming, the Horn of Africa is experiencing prolonged and severe drought conditions, threatening the lives of both people and animals. In recent years, rain has become less predictable and the wet season has shortened, meaning it is more difficult to grow food, provide for animals and maintain the environment. The conditions have been worsening over time. Between 1979 and 2000, rainfall measurements were several hundred millimetres lower than historical averages and by 2006–2009 it was reported that vegetation conditions were 50 to 100 per cent below average conditions. By 2011, the long-term lack of rainfall and vegetation resulted in a humanitarian crisis where more than 100 000 people died as a result of severe famine.

Source 4

Food security in the Horn of Africa.

Country	Population without food security	Total population
Ethiopia	5.6 million	112 million
Somalia	2.9 million	15 million
Kenya	2.7 million	52 million
Uganda	1.6 million	44 million

Animals and plants native to the region are also affected by the drought; however, livestock and domestic animals that are not adapted to prolonged hot conditions tend to be more severely affected. In 2016, the Food and Agriculture Organization of the United Nations (FAO) predicted that tens of thousands of livestock had died or become sick as a result of the drought. In 2017, many countries in the Horn of Africa were again identified by FAO as places of significant concern and a 'pre-famine alert' was issued. The FAO reports that food insecurity has almost doubled as a direct result of the drought, with over 12.8 million people in need from Kenya, Somalia, Ethiopia and Uganda alone.

Source 5

Drought and high average annual temperatures mean that water security is also low. People may need to walk more than 6 kilometres every day to fetch water.

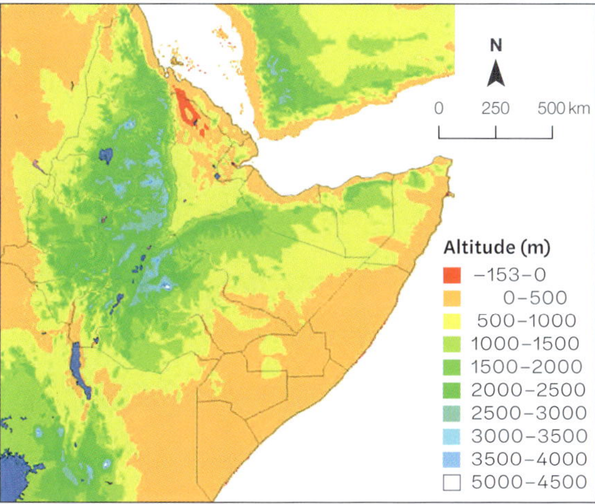

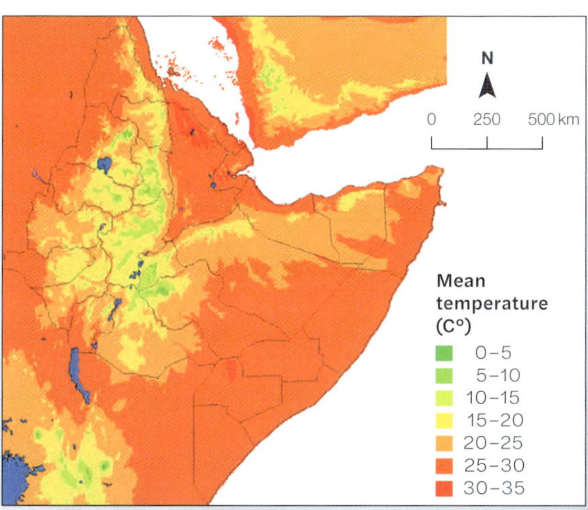

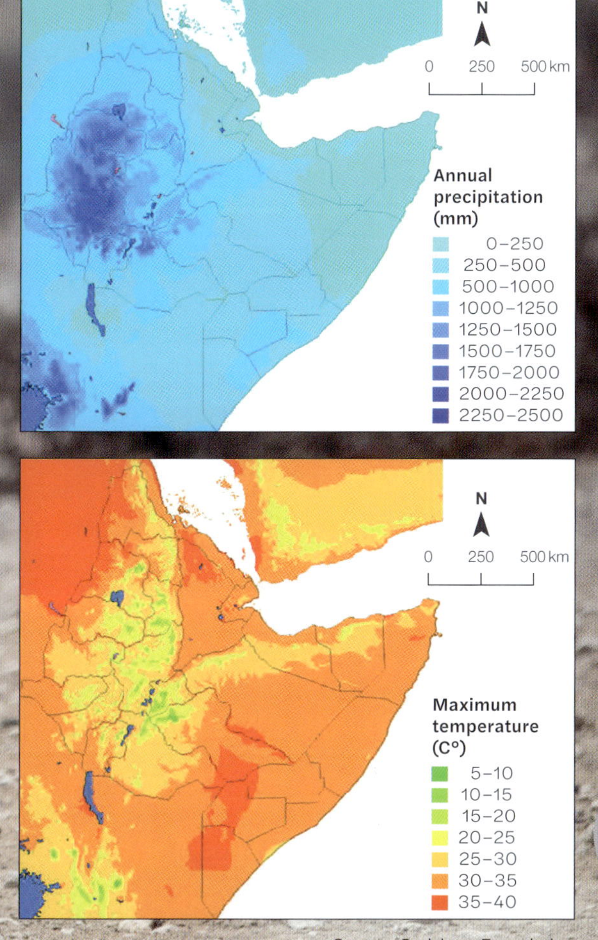

Source: Carlsberg Foundation

Source 6

Climate (precipitation and temperature) in the Horn of Africa

Learning ladder G2.3

Show what you know

1 Compare the terms 'wildfire' and 'bushfire'.

2 As a class discuss the following statement: 'Droughts and bushfires are not associated with changing climates'.

3 Use a mind map or a diagram to illustrate how extreme weather events such as fire or drought can lead to environmental change.

Digital and geospatial technologies

Step 1: I can interpret different map types using cartographic conventions

4 Discuss how different maps can be used to manage and assess the environmental impact of natural disasters.

Step 2: I can construct paper maps using correct cartographic conventions

5 Source 4: Using the data from the table, construct a map of the Horn of Africa, illustrating food security variation in the *region*.

6 Sources 5 and 6: As a pair, describe the *spatial association* between climate and food security.

Step 3: I can access and use geospatial technology platforms such as GIS

7 Access http://mea.digital/GHV10_G2_1. Outline how this technology would be useful in assessing environmental *change* and human management over time.

Step 4: I can manipulate data using digital and geospatial technologies

8 Access http://mea.digital/GHV10_G2_2. Explore the interactive map and note areas that are currently affected by fire. What are the characteristics of these places? How might the risk of fire increase or decrease over time?

HOW TO

Mapping with BOLTSS, page 152
PQE, page 154
SHEEPT, page 156
Sketches and annotating, page 158
Answering questions, page 162

What is the importance of the Belo Monte Hydroelectric Dam?

Brazil has long been considered a sleeping economic giant – a populous country with vast natural resources. In recent decades, increasing rural–urban migration has placed pressure on Brazilian cities, causing rising demand for energy to promote economic growth. The development of the Belo Monte Hydroelectric Dam has the ability to solve this issue, but is the environmental cost worth it?

Located in the northern Brazilian state of Pará, in the Amazon Basin, the Belo Monte Hydroelectric Dam was a polarising project. On the one hand, it offered significant potential for economic growth and could supply power to up to 23 million homes; on the other hand, it was a potentially devastating threat to biodiversity and Indigenous lands.

Completed in November 2019, the dam can produce up to 11 200 megawatts of power, but it has generated significant controversy since it was first proposed in the 1970s. At the 1989 Altamira Gathering, a First Nations Kayapó woman, Tuíra, famously threatened the chief engineer with a machete, an image that inspired global outrage at the dam's construction. Opposition to the project grew, forcing the World Bank to cancel a proposed $500 million loan to Brazil. Nevertheless, in time the project again gathered momentum and, despite continued public opposition and legal challenges, was successful.

‘We do not need electricity. Electricity is not going to give us our food. We need our rivers to flow freely: our future depends on it. We need our jungles for hunting and gathering. We do not need your dam.’

Source 1

In Kayapó, her native tongue, Tuíra delivered a message for the chief engineer.

Source 2

Indigenous peoples across Brazil widely protested the construction of the dam.

Impacts of the dam

- At least 400 km^2 of the Amazon Rainforest flooded by the dam
- Disruption to traditional water flows, frequencies and volume, as well as to tributaries up to 130 kilometres downstream
- Loss of habitats and biodiversity; the region contains more than 600 endemic species of fish and other rare flora and fauna, such as the Xingu Poison Dart Frog (*allobates crombiei*)
- Significant amounts of greenhouse gases emitted, as foliage decays and releases methane
- More than 20 000 people displaced, especially traditional owners
- Population growth significant during construction; from 2011–2014, the population of the Altamira region doubled from 75 000 to 150 000
- Creation of almost 20 000 direct jobs and 25 000 indirect jobs
- A rise in social issues in the Altamira region, which now has the highest homicide rate in Brazil
- Over the next 50 years alone, energy produced by the dam will be the equivalent of more than 1.1 billion barrels of oil

Canadian mining company Belo Sun also plans to open the Volta Grande Project, Brazil's largest gold mine. Forecast to produce around $200 million worth of gold over two decades, the project has met with firm opposition from First Nations groups and environmentalists, who are concerned about the impact of the mining on the traditional lands of the Juruna and Arara peoples. There is also concern about the use of dangerous chemicals, such as cyanide, and the proper disposal of **tailings** and other waste. In many parts of the world, such as in Papua New Guinea, the leaching of these materials into the local ecosystems has caused significant environmental damage.

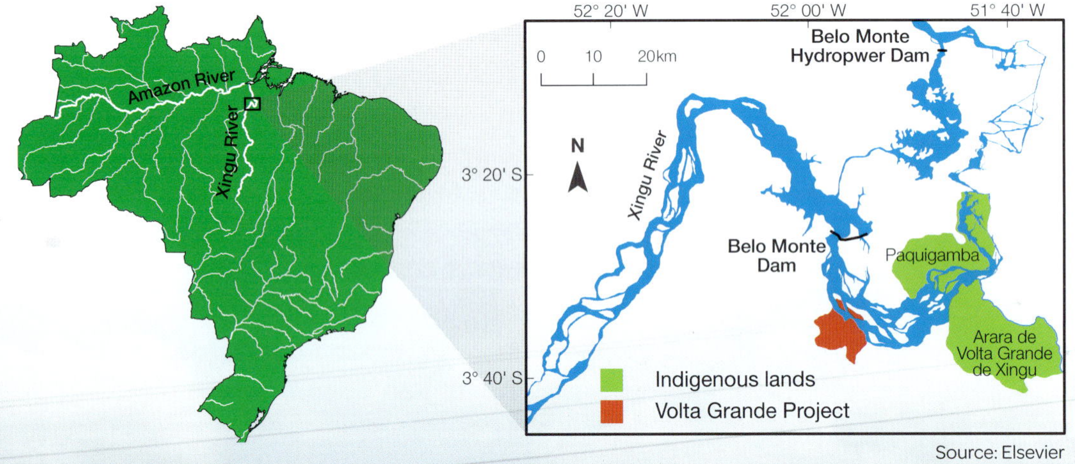

Source: Elsevier

Source 3

The Hydroelectric Dam complex and the proposed Volta Grande Project are both just a short distance from Indigenous lands in north-central Brazil.

Source 4

The Monte Belo Hydroelectric Dam took decades to gain final approval and years to build.

Source 5

Raoni Metuktire is a leader of Kayapó people of the Amazon Basin and has been an ardent campaigner against the construction of the Belo Monte dam.

Learning ladder G2.4

Show what you know

1 Why was the Belo Monte Hydroelectric Dam built?

2 Tuíra and Raoni Metuktire have been prominent campaigners against the dam. Suggest why they are opposed to it, despite the economic benefits it may bring.

Changes and implications

Step 1: I can identify that changes occur in the characteristics of places over time

3 List three specific positive and three specific negative impacts of the Belo Monte Hydroelectric Dam.

4 Rank your selections in terms of the most positive and most negative and, with a partner or in small groups, justify your rankings.

Step 2: I can describe how places have changed over time

5 Building the dam required vast resources and generated many jobs. How would this have affected infrastructure, housing and other industries in the area?

Step 3: I can explain the causes behind the change over time in a place

6 Using *interconnection*, explain how the construction of the dam will negatively affect biodiversity in the *region*.

7 Explain how the dam will affect the future *sustainability* of the traditional ways of life of the First Nations peoples of the *region*.

Step 4: I can make predictions and outline consequences of change over time

8 Conduct further research into the Volta Grande Project, particularly the 'Belo Sun No!' protest campaign, and then draft a letter to the editor, outlining a strong argument for or against the mine. You could start with http://mea.digital/GHV10_G2_3.

HOW TO

PQE, page 154
SHEEPT, page 156
Answering questions, page 162

How does human management impact place?

The impacts of human activities on natural landscapes are profound and act as agents of significant change. Historically, environmental management has often been poorly understood and short-term economic gains have taken priority, particularly in the contexts of growing populations and economic development. The consequences of mismanagement can have devastating environmental and social impacts on places and communities. Both the Aral Sea of Central Asia and Australia's Murray–Darling Basin serve as stark examples of the impacts of ineffective human management.

The Aral Sea

Across the borders of Kazakhstan and Uzbekistan, the Aral Sea was once the fourth-largest lake in the world at 68 000 km^2. Fed by the inflows of two of the largest rivers in Central Asia, the Syr Darya and Amu Darya, it was a valuable source of fresh water, fishing and transport in a relatively arid region.

During Soviet times, the government began to divert water away from the rivers to feed irrigation projects, primarily to produce cotton in the fertile Ferghana Valley. This formed part of a broader campaign to transform the geography of the nation (Source 2). Poor irrigation infrastructure meant that, in some cases, up to 70 per cent of the diverted water was lost through evaporation, infiltration and leakage.

Source 1

An abandoned ship in the Aral Sea

Reduced flow into the Aral Sea has had a devastating effect. By 1997, the lake covered just 10 per cent of its original size. The smaller concentration has led to alarmingly high levels of salinity, with the water now 10 times saltier than sea water. Biodiversity has been diminished as migratory birds no longer return and local species dwindle. Pesticides, chemicals and salt that remain in the dried-out areas are now carried by wind beyond the immediate area and into the broader region, causing increases in cancers, tuberculosis, anaemia and other illnesses, along with a rise in infant and maternal mortality rates, particularly among the Karakalpak people of northern Uzbekistan. Toxins from the pesticides have been found in the breast milk and blood of mothers. Birth abnormalities in the region are five times higher than average.

The Aral Sea showing original distribution and current-day extent.

Source: shutterstock.com

Source 3

The North Aral Sea and South Aral Sea are all that remain of the original Aral Sea itself.

Source 2

A 1949 Soviet poster for the Great Plan for the Transformation of Nature campaign. The text on the poster means, 'And defeat drought!'.

Limited evaporation has also affected the water cycle in the area, triggering droughts, which only further exacerbates issues with dust storms and wind erosion. The fishing industry, which once supplied one-sixth of the Soviet Union's total fish catch, is now abandoned and unemployment and economic hardship are widespread; the fishing towns that dotted the historical shoreline in Uzbekistan are now more than 150 kms away from the water.

The Murray–Darling Basin

The Murray–Darling Basin is a water catchment area in south-east Australia, encompassing both the Murray and Darling rivers and their tributaries. The area is home to more than 2.2 million people, 9200 irrigated agriculture businesses and dozens of endemic species of flora and fauna. On average, 24 000 gigalitres of water flows through the basin per year, although generally less than half this amount – and considerably less during dry periods – reaches the sea at South Australia. This makes the basin one of the world's lowest in terms of volume flow.

The semi-arid conditions of much of the region, combined with variability of climate, play a significant role in this lack of volume, as evaporation levels are high and limited rainfall keeps the flow quite slow for most of the year. The basin itself also contains more than 30 000 wetlands, including many of international significance, which rely on the ebb and flow of the river for their ongoing health. For tens of thousands of years, the First Nations peoples of the region have lived in harmony with the land, adapting to its natural rhythms and variations.

The Murray–Darling Basin

Source 5

The Murray–Darling Basin extends from Queensland in the north, via NSW and Victoria to South Australia.

Source: Murray–Darling Basin Authority

Source 4

The Darling River

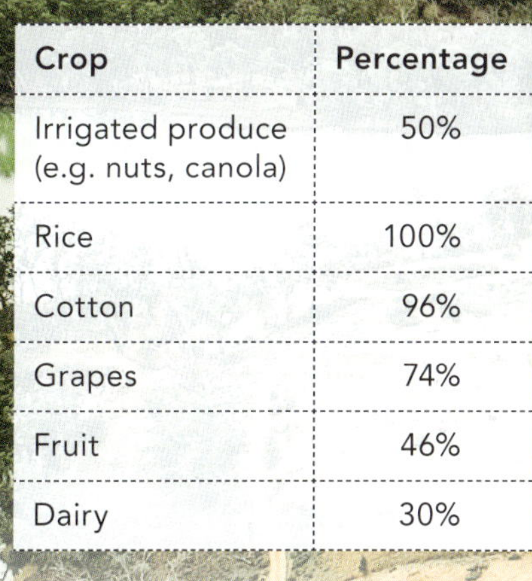

Crop	Percentage
Irrigated produce (e.g. nuts, canola)	50%
Rice	100%
Cotton	96%
Grapes	74%
Fruit	46%
Dairy	30%

Source 6

The percentage of Australia's crops and dairy produced in the basin

Source 7

Low water levels in the Murray River

The arrival of Europeans into the region in the 19th century had a dramatic effect on the basin. Using agricultural techniques suited to temperate lands, British farmers set about diverting water for irrigation projects and introduced foreign species of flora and fauna. As the population grew, the pressure on the delicate ecosystem increased and natural variability of precipitation pushed water supplies to the brink. At the other extreme, flooding caused significant challenges, so authorities set about establishing locks, dams, coastal barrages and weirs as mechanisms to control the flow of water. These changes have had a tremendous effect on the system, as 11 500 gigalitres of water is diverted away from the basin for agriculture, industry and domestic use.

The construction of the Snowy River Hydroelectric Project, completed in 1974, which diverted water from outside the basin into it via an elaborate system of tunnels and turbines has ensured an extra 2500 gigalitres of water have been added into the system. This has provided both precious natural flow and water for irrigation, supporting the billions of dollars of direct and indirect economic benefits it provides.

The health of the Murray–Darling Basin is in the best interest of all Australians, but the finite water resources are increasingly under pressure from the competing needs of users and the environment. Careful environmental management is crucial for the future health of the system, particularly as the impacts of global warming continue to add uncertainty to the region.

Learning ladder G2.5

Show what you know

1 Why was water diverted away from the Aral Sea?

2 Initially, Australian environmental management decisions were not made in consultation with First Nations peoples. Why might this have been the case? How could things have been different?

Changes and implications

Step 1: I can identify that changes occur in the characteristics of places over time

3 Create a table that identifies environmental changes as a result of poor management at both the Aral Sea and Murray–Darling Basin.

Step 2: I can describe how places have changed over time

4 Using a flow chart, outline how the *process* of diverting water away from a river affects its health.

Step 3: I can explain the causes behind the change over time in a place

5 Agriculture in both Australia and Uzbekistan is a multibillion-dollar industry and populations continue to grow. Explain why socioeconomic needs in both countries have taken priority over environmental needs.

Step 4: I can make predictions and outline consequences of change over time

6 You are the new Director of Projects for the national water board. With reference to either the Aral Sea or the Murray–Darling Basin, describe how you will improve the situation for both people and the *environment*, while considering how those who benefit from water diversion might also be adversely affected.

SHEEPT, page 156
Answering questions, page 162

How do governments attempt to balance sustainable development?

The mining industry can be worth billions of dollars in export revenue and can create employment and support infrastructure development. However, mining can also cause damage to the environment and threaten the lands and cultural heritage of Indigenous peoples.

Balancing the social, economic and environmental elements of mining is the key to sustainable development. We can study the difficulties of this balance by looking at the Ok Tedi and Carmichael mines.

Ok Tedi

Located near the headwaters of the Ok Tedi River, close to Papua New Guinea's (PNG) western border, the Ok Tedi mine is a large open-cut copper and gold mine. The mine has been a boon to the national economy, responsible for more than a quarter of the country's entire export earnings, producing almost 10 million tonnes of copper and almost 50 tonnes of gold bullion.

The mine has created jobs and helped fund nearby infrastructure. The town of Tabubil was built to support employees and includes a school and medical facilities, along with the Kiunga-Tabubil Highway. Kiunga, a port town of almost 10 000 people, also benefits by providing haulage, shipping and transport of mined materials and other resources.

However, the mine has also left a legacy of environmental destruction. Up until 2006, the mine operators discharged more than 80 million tonnes of waste into the local river system. This toxic waste poisoned the local fish population, and the added silt raised riverbeds, increasing flooding. When flooding subsided, cultivated fields adjacent to the rivers downstream were covered with toxic mud. More than 50 000 people living downstream of the mine faced food shortages and dangerously high concentrations of copper and other elements in their water supply. Sites of traditional significance were lost and coral reefs at the mouth of the river were damaged.

Carmichael

The Carmichael coal mine, located in the Galilee Basin in central Queensland, remains a controversial issue in Australia. Owned by the Adani Group, people have protested against it, amid concerns about its effect on the environment. The protests have even led to the Adani group changing its name to 'Bravus' to avoid negative publicity.

Government support for the mine has centred around economic benefits: the mine is forecast to generate almost $3 billion annually for Queensland with the potential to create around 1500 direct jobs, plus around 6750 indirect jobs. The high-quality coal is to be transported to Abbott Point coal port and will require the construction of rail infrastructure and new shipping terminals, as well as dredging of the port itself. Given the forecasted economic benefits, the mine has support from both the federal and Queensland governments. However, government subsidies to support its development have also been controversial.

The mine has become a lightning rod for protests about climate change and the need for the nation to reduce carbon emissions. The potential amount of carbon dioxide emissions coal from the mine will produce is around 200 million tonnes over its 60-year life.

The mine will use groundwater and water from the Belyando River, diverting precious water resources in a drought-prone region. The proposed site is also home to a number of endangered species.

Papua New Guinea

Source: Alamy.com/Custom Mapping Services

Queensland's coal basins

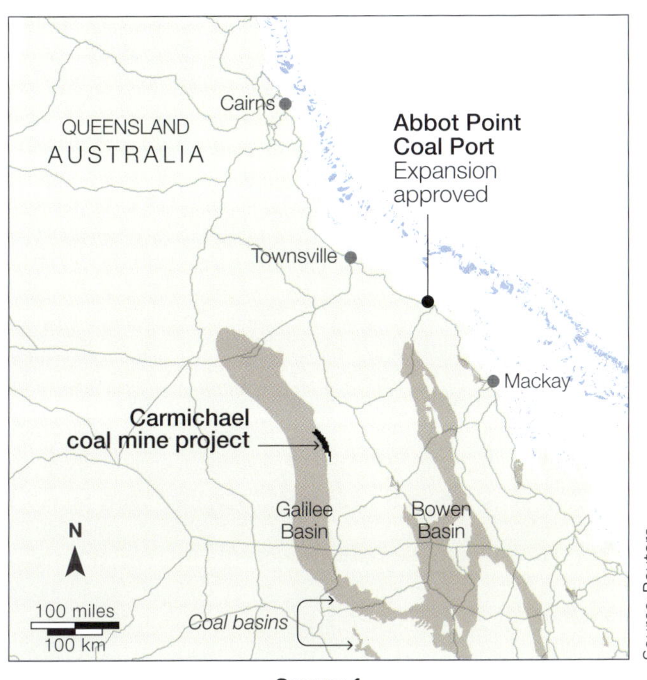

Source 1

The Ok Tedi and Carmichael mines

The final major concern has been the impact on the traditional owners of the land, the Wangan and Jagalingou peoples, with whom there has been little consultation and whose Native Title challenges have not been successful.

Learning ladder G2.6

Show what you know

1 Describe the relative location of the Ok Tedi mine.

2 Why is Adani so prominently featured at climate rallies around Australia?

Changes and implications

Step 1: I can identify that changes occur in the characteristics of places over time

3 List five positive direct or indirect impacts of the Ok Tedi mine.

Step 2: I can describe how places have changed over time

4 Select a specific negative impact from both Ok Tedi and Carmichael. For each, describe why they would be considered negative.

Step 3: I can explain the causes behind the change over time in a place

5 With reference to either Ok Tedi or Carmichael:
  a explain why the mine is good for the economy and society.
  b explain why the mine is bad for the *environment* and society.
  c suggest why there is a genuine need for governments to balance the sustainable development of such a project.

Step 4: I can make predictions and outline consequences of change over time

6 Prepare a SWOT analysis of the Ok Tedi mine.

7 a Add further points to your analysis with information from http://mea.digital/GHV10_G2_4.
  b Using your analysis, evaluate the mine for the PNG government.

HOW TO
SHEEPT, page 156
Answering questions, page 162

Getting involved

How can I make a difference?

Recycling should be part of our everyday life, whether it be using the right bin to dispose of paper waste at school, composting food scraps from the kitchen or buying second hand clothes from op-shops. However, the amount of waste Australia produces is still increasing, with around 75.8 million tonnes of rubbish thrown away in 2018–2019 alone, a 10 per cent increase since 2016–17.

In some states, financial recycling incentives, such as the 10-cent recycling refunds on drink cans and bottles in South Australia, have been successful in reducing landfill. While interest in buying and selling second hand is slowly increasing, large items such as furniture still tend to end up on council hard rubbish piles.

In October, 2019, Swedish store Ikea launched a buyback scheme, offering customers who returned used Ikea furniture store vouchers worth up to half the original price, depending on quality and condition. The international **incentive**, running in 27 countries, allows customers to log their item online to estimate the return value. Ikea then resells items or recycles them for customers. Ikea aims to be 'a fully circular and climate positive business by 2030', meaning it both sells and recycles its own products.

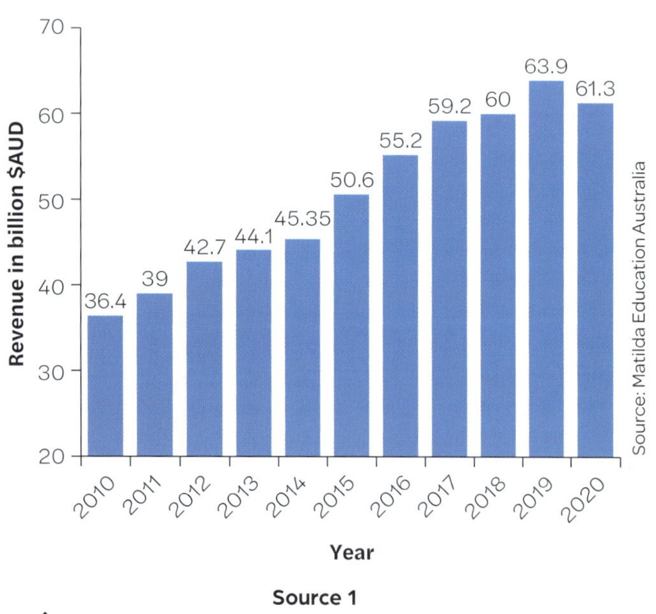

Source 1

IKEA was the leading furniture retailer in Europe by far in 2019 based on revenue.

Source 2

A promotional image for Ikea circular living – what do they mean by circular living?

Since the 10 Australian Ikea stores began the incentive in 2019, over 10 000 Ikea products have been returned, equating to over 100 tonnes of recycled furniture. Further, the scheme is allowing cheaper furniture to be available for consumers, with Ikea reselling over 47 million recovered products last year.

Ikea's buyback scheme is one example of ways you could make a change and reduce your carbon footprint. Remember, many small actions can lead to large impacts. You only need one person to start a change.

Source 3

An Ikea store

Getting involved

Ikea's circular program is just one example of how we can recycle products and reduce our carbon footprint.

1 Brainstorm other creative ways large companies or industries could be more sustainable when producing and managing their products.
2 Complete an audit of some of the rubbish that your school or home produces.
   a Create a table highlighting key rubbish you expect to find in your school grounds; e.g. plastic, paper, aluminium foil.
   b Assign groups of two or three to different sections of the school.
   c Complete a tally of how much rubbish you find during a 10-minute survey.
   d Back in the classroom, graph your results using a bar graph.
   e Compare your results with other groups. Were there some areas of the school that had more rubbish than others? Why? Was there one particular type of rubbish that was more prevalent than others? Why?
   f Based on your class results, review how much of the rubbish found could actually be recycled, either through direct reuse or by being made into something else.
   g Create a secondary tally of how many recyclable products are being thrown away in this *place*.
   h Write a short 200–300-word reflection addressed to your principal that highlights your main results and ideas about how to reduce your school's waste output.
3 Not all furniture comes from Ikea! Create a series of posters that makes your local community more aware of how to recycle or responsibly dispose of unwanted items. Some sites to find more information include:
   - http://mea.digital/GHV10_G2_6
   - http://mea.digital/GHV10_G2_7.

Masterclass

Learning ladder

How has deforestation changed Borneo's environment?

Geography is the study of people and place and how different processes and phenomena interact. Throughout G1 you explored the causes and consequences of environmental change and then in G2 you have focused on specific examples of this change. In Geography, a key skill is being able to apply information and ideas that you have learned from other scenarios to explore and make sense of new case studies.

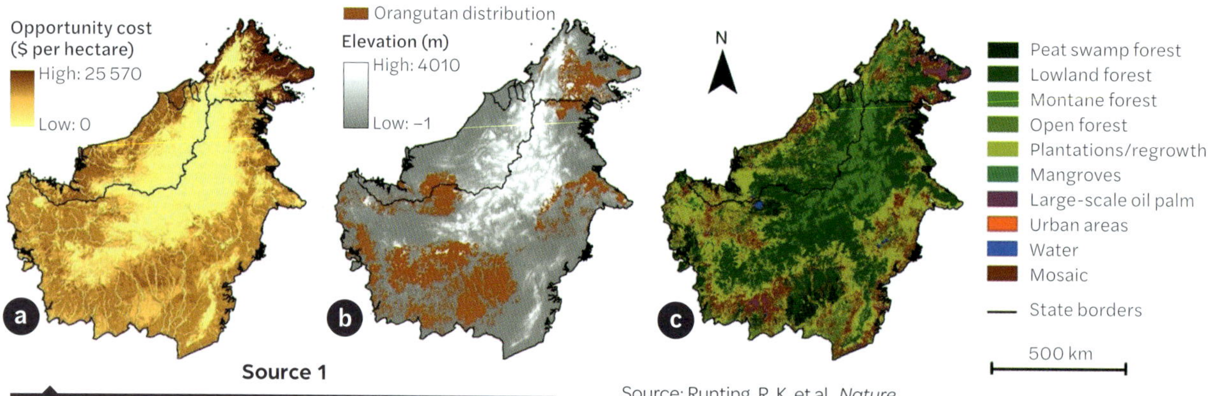

Source 1

Maps of Borneo representing different elements affected by palm oil production: **a** opportunity cost, **b** orangutan distribution and **c** the current vegetation distribution.

Source: Runting, R. K. et al., *Nature*

Oil palm production

Oil palm crop production is measured in tonnes.

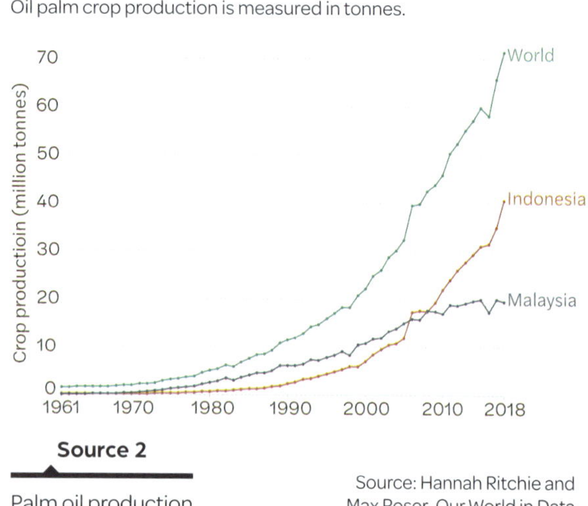

Source 2

Palm oil production

Source: Hannah Ritchie and Max Roser, Our World in Data

Oil yield by crop type, 2017

Global oil yields are measured as the average amount of vegetable oil produced per hectare of land. This is different from the total yield of the crop since only a fraction is available as vegetable oil.

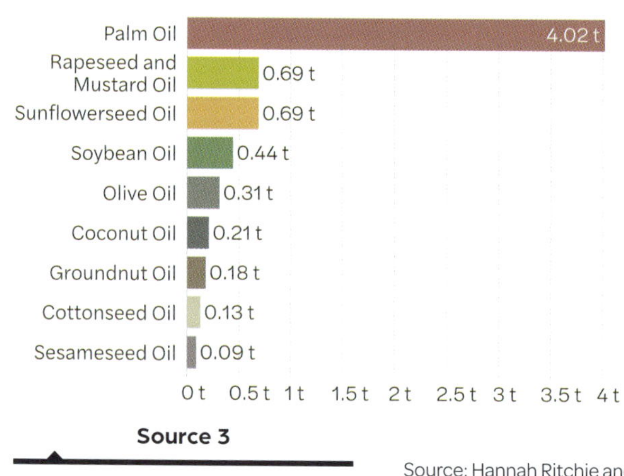

Source 3

Yields per hectare of different crops

Source: Hannah Ritchie and Max Roser, Our World in Data

Step 1

a I can identify spatial distributions and patterns

Source 1: Using Google Maps, describe the relative location of Borneo.

b I can provide short explanations for patterns and interconnections

Suggest changes to forests that have large-scale consequences for global climate. (Hint: refer to G1.7 for background information.)

c I can identify that changes occur in the characteristics of places over time

Source 2: Describe how the quantity of palm oil production has changed over time in Indonesia.

d I can list primary and secondary methods useful for my study

Source 1: Suggest how orangutan numbers may have been estimated to create map **b**.

e I can interpret different map types using cartographic conventions

Suggest two reasons why geospatial technology is useful in exploring changes to *environments* on different *scales*.

Step 2

a I can use data to quantify spatial distributions and patterns

Source 3: Rank crops according to their efficiency per hectare.

b I can explain patterns and interconnections

Source 1: Using SHEEPT, explain why the orangutan *distribution* is so limited in Borneo.

c I can describe how places have changed over time

Source 2: Outline how an increase in palm oil production in Malaysia and Indonesia may have led to environmental change in the *region* of Borneo (which is part of both countries).

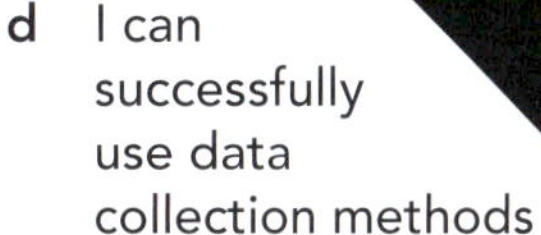

d I can successfully use data collection methods

Using reliable sources, collect 10 quantitative statistics that help describe environmental change in Borneo.

e I can construct paper maps using correct cartographic conventions

Using a blank world map, find Borneo and state the absolute location of the capital.

Step 3

a I can describe spatial distributions and patterns

Source 1: Using PQE, describe the *distribution* of land cover in Borneo.

b I can use data to support explanations of patterns and interconnections

'Borneo is an example of how humans negatively impact the *environment*.' To what extent do you agree with this statement? Use data from research in your response.

c I can explain the causes behind the change over time in a place

Source 2: Explain why palm oil production may have increased over time.

d I can filter collected data

Create a data table that describes the social, economic and political *environment* in Borneo.

e I can access and use geospatial technology platforms such as GIS

Access http://mea.digital/GHV10_G2_5. Explore and list the data available. Suggest why this data is available for public access.

Masterclass

Step 4

a I can use data to support exceptions to spatial distributions and patterns

'Deforestation has only negative impacts on people and place.' Do you agree with this statement? Use data to show some positive impacts of deforestation.

b I can use relevant sources to research further reasons for patterns and interconnections

Discuss how Borneo faces a conflict between environmental and economic *sustainability*. (Hint: refer to G1.14 for background information.)

c I can make predictions and outline consequences of change over time

Predict how varied world views and opinions may influence future environmental *change* in Borneo. (Hint: refer to G1.13 for background information.)

d I can organise data collected according to relevance for a research question

Discuss how geospatial technology may help governments and policymakers manage environmental change in Borneo. (Hint: refer to G1.3 for background information.)

e I can manipulate data using digital and geospatial technologies

Access http://mea.digital/GHV10_G2_5 again. Select one area of interest such as land cover changes, uses of protected areas or changes in concessions (permits). Turn layers on and off and write a short paragraph highlighting any patterns or *distributions* that help explain environmental change in this place.

Step 5

a I can identify multiple spatial distributions and patterns

Source 1: Describe the spatial association between land cover and opportunity cost in Borneo.

b I can interpret causes of patterns and interconnections

Source 1: Explain the *interconnection* between opportunity cost and orangutan *distribution* in Borneo.

c I can interpret data to quantify predictions based on research

Go to Source 1 on page 44: Based on temperature mapping, predict how global climate change may impact places such as Borneo in the future.

d I can evaluate the success of research methods

Sources 1 and 2: Outline the reliability of the data presented. Consider source, data type and year of collection.

e I can draw conclusions from geographical information in digital and geospatial technologies

Research the current management strategies imposed by Indonesian and Malaysian governments for managing Borneo's *sustainability*. Evaluate the effectiveness of one strategy.

Capstone

How can I understand place studies of change and management?

In this chapter, you have learned a lot about change and management. Now you can put your new knowledge and understanding together for the capstone project to show what you know and what you think.

In the world of building, a capstone is an element that tops off a building or a wall. That is what the capstone project will offer you, too: a chance to top off and bring together your learning in interesting, critical and creative ways. You can complete this project yourself, or your teacher can make it a class task or a homework task.

mea.digital/GHV10_G2

Scan this QR code to find the capstone project online.

Development and disparity

G3

How can I understand human wellbeing?

Human wellbeing is an expression of health, safety, happiness and other tangible and intangible measures that vary between people and places. However, human wellbeing has no globally recognised definition, as it is a complex concept. It involves many aspects of our lives including social, historical, economic, environmental and political factors.

learning ladder

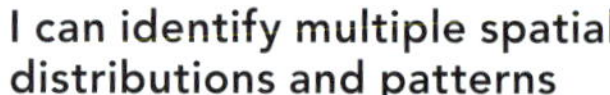

Step	Spatial distributions and patterns	Patterns and interconnections	Changes and implications
step 5	**I can identify multiple spatial distributions and patterns** I can take my PQE one step further to find links or relationships that exist in development and disparity.	**I can interpret causes of patterns and interconnections** I can use multiple sources to find links or relationships that exist in development and disparity and can explain 'Why?'.	**I can interpret data to quantify predictions based on research** I can use external data from research as evidence of the positive and negative impacts of a change I have predicted.
step 4	**I can use data to support exceptions to spatial distributions and patterns** I can use data to answer 'Why?' about the exceptions identified in a PQE analysis of development and disparity.	**I can use relevant sources to research further reasons for patterns and interconnections** I can use sources other than this textbook to further research patterns I observe in development and disparity.	**I can make predictions and outline consequences of change over time** I can use my knowledge of natural processes and world regions to make an educated guess about the positive and negative impacts of change in development and disparity.
step 3	**I can describe spatial distributions and patterns** I can describe patterns, quantify them and point out exceptions (PQE) to describe development and disparity.	**I can use data to support explanations of patterns and interconnections** I can use data from a map or graph to explain patterns I observe in development and disparity.	**I can explain the causes behind the change over time in a place** I can use my knowledge of natural processes and world regions to explain why changes may occur over time in development and disparity.
step 2	**I can use data to quantify spatial distributions and patterns** I can read data and use it to measure key trends on a map or graph about development and disparity.	**I can explain patterns and interconnections** I can identify social, historical, economic, environmental, political and technological (SHEEPT) factors to help me explain development and disparity.	**I can describe how places have changed over time** I can use specific examples to describe changes over time in development and disparity.
step 1	**I can identify spatial distributions and patterns** I can find key trends on a map or graph about development and disparity.	**I can provide short explanations for patterns and interconnections** I can write descriptions of patterns and interconnections that I find in development and disparity.	**I can identify that changes occur in the characteristics of places over time** I can read information and answer questions about changes over time in development and disparity.

Warm up

Source 1

Safety and peace are key elements to human wellbeing and happiness.

I can evaluate the success of research methods
I can look back and comment on the data collection methods I used, and evaluate how successful they were in helping me answer a research question about development and disparity.

I can organise data collected according to relevance for a research question
I can review the data I have collected in the field and display it using graphs, tables, annotations and captions.

I can filter collected data
I can review my collected data and select the most relevant data to answer a research question about development and disparity.

I can successfully use data collection methods
I can use primary and secondary data collection methods in the field and classroom to investigate development and disparity.

I can list primary and secondary methods useful for my study
I can create a checklist of methods to investigate development and disparity and categorise them as primary or secondary methods.

Communicate data

I can draw conclusions from geographical information in digital and geospatial technologies
I can interpret and analyse patterns by using different layers and features on geospatial technology platforms.

I can manipulate data using digital and geospatial technologies
I can work with layers and other features on geospatial technology platforms to further explore data and interconnections.

I can access and use geospatial technology platforms such as GIS
I can use geospatial technology platforms to explore data and find patterns.

I can construct paper maps using correct cartographic conventions
I can use a pencil, paper and ruler to construct a map that follows BOLTSS conventions.

I can interpret different map types using cartographic conventions
I understand data found in different types of maps and graphs and use the data to answer questions about development and disparity.

Digital and geospatial technologies

Spatial distributions and patterns

1 Predict how human wellbeing will vary on a global *scale*.

Patterns and interconnections

2 Suggest how wellbeing is interconnected with our daily happiness.

Changes and implications

3 Outline how crises such as COVID-19 may impact people's wellbeing in your community.

Communicate data

4 Describe how the situation illustrated in Source 1 may influence human wellbeing.

Digital and geospatial technologies

5 List two data sources that would be useful in developing a map to visualise economic inequality on a global *scale*.

What is human wellbeing?

Human wellbeing is centred on the idea that people want to be healthy, happy and safe. Human wellbeing is multidimensional; that is, a range of factors influence wellbeing depending on person and place. Wellbeing can be affected by a person's access to social connection, income, food and resources, peace, social and economic equality, and freedom. Wellbeing is also influenced by a person's feelings, emotions, thoughts and life satisfaction. The way a person describes their wellbeing may change over time as their circumstances change.

Happiness vs wellbeing

The terms happiness and wellbeing are often used interchangeably. Happiness is often associated with a positive wellbeing and is generally centred on comparisons people make with others. Happiness can fluctuate from day to day and is measured or felt differently by different people. You may be happier today than yesterday because the weather is sunny and you can go to the beach, or you may be happier than your brother because you got the last biscuit in the pack. The feeling of happiness is also often linked with memories or nostalgia.

When we discuss wellbeing, however, we are considering the ongoing assessment of people's lives. We examine social, economic, environmental and political factors. These may include:

- access to reliable water and food
- stable employment and occupational safety
- exposure to or involvement in war or conflict (including domestic abuse)
- physical and mental health
- community connections and relationships
- access to education, healthcare and infrastructure.

All of these factors contribute to a person's wellbeing as well as their day-to-day happiness.

Source 1

A range of factors determine a person's wellbeing over their lifetime.

Source 2

Happiness and wellbeing are determined by a range of factors including local climate, living conditions, safety and freedoms. Ahmed, from Kismayo in Somalia, lives at the Dadaab refugee camp in Kenya. He is a political refugee and receives a ration card and a tent to live in. However, like thousands of other refugees, he is not allowed to live or work outside the refugee camp. His situation has a strong impact on his wellbeing.

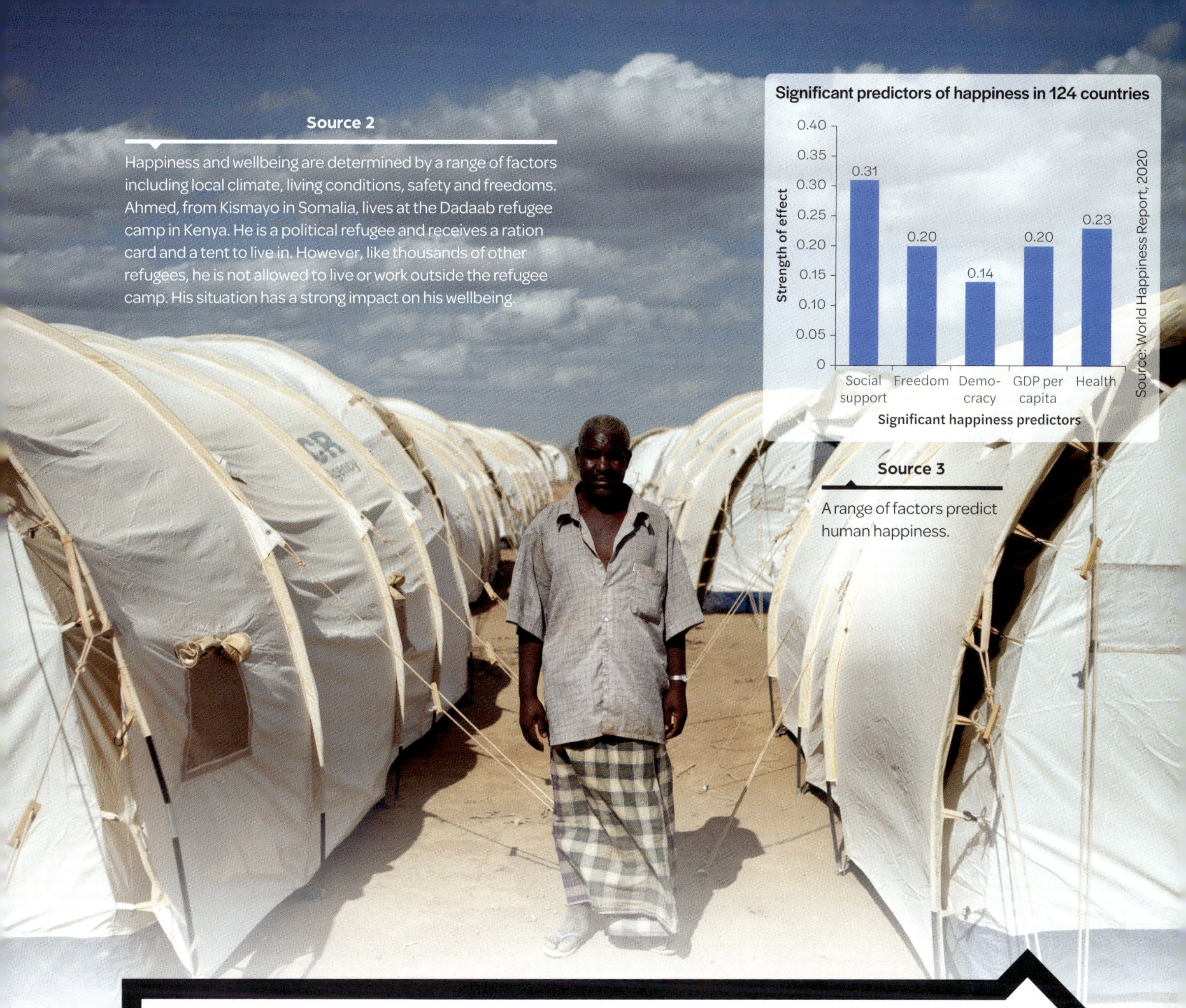

Significant predictors of happiness in 124 countries

Strength of effect
0.40
0.35
0.30
0.25
0.20
0.15
0.10
0.05
0
0.31
0.20
0.14
0.20
0.23
Social support
Freedom
Democracy
GDP per capita
Health
Significant happiness predictors
Source: World Happiness Report, 2020

Source 3

A range of factors predict human happiness.

Learning ladder G3.1

Show what you know

1 Create a mind map highlighting factors in your life that you associate with day-to-day happiness.

2 Summarise the *interconnection* of wellbeing and happiness.

3 Suggest why happiness and wellbeing may differ between people and *place*.

Communicate data

Step 1: I can list primary and secondary methods useful for my study

4 Imagine you are part of a research group measuring happiness and wellbeing in Australian teenagers. List two **primary methods** and one **secondary method** you could use to gain data to answer your research question.

Step 2: I can successfully use data collection methods

5 Select one primary research method you outlined in Question 4 and carry it out among your classmates.

Step 3: I can filter collected data

6 Using reliable resources, research what other people around the world associate with happiness and wellbeing.

Step 4: I can organise data collected according to relevance for a research question

7 Create a photo essay that illustrates varying indicators and contributing factors to wellbeing and happiness on various *scales*.

Photo essays, page 160
Answering questions, page 162

How is wellbeing distributed on a global scale?

Around the world, human wellbeing and happiness are unevenly distributed. We can largely attribute this to the uneven distribution of resources between regions.

Happiness around the world, according to the Happy Planet Index

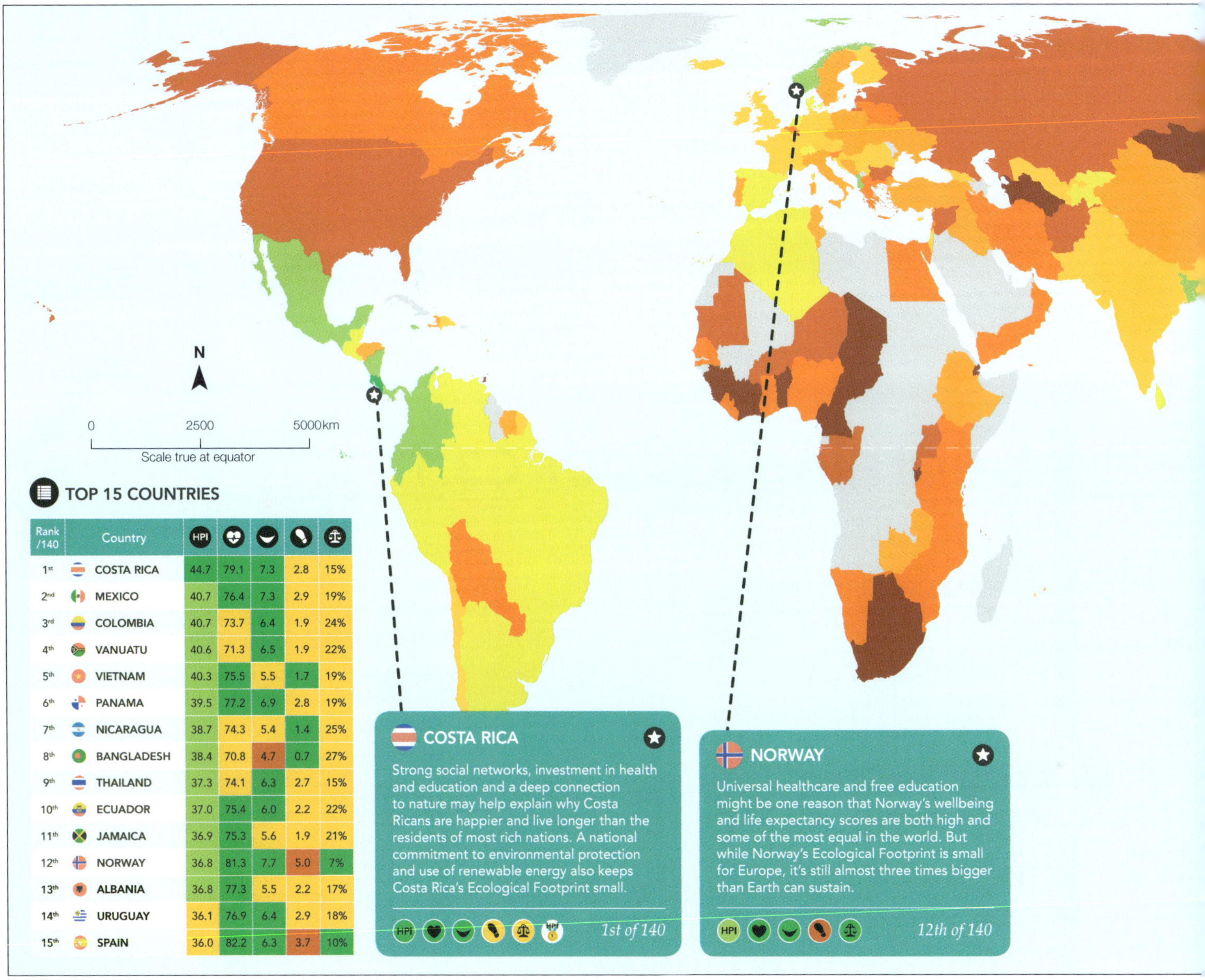

N

0 2500 5000km

Scale true at equator

TOP 15 COUNTRIES

Rank /140	Country	HPI				
1st	COSTA RICA	44.7	79.1	7.3	2.8	15%
2nd	MEXICO	40.7	76.4	7.3	2.9	19%
3rd	COLOMBIA	40.7	73.7	6.4	1.9	24%
4th	VANUATU	40.6	71.3	6.5	1.9	22%
5th	VIETNAM	40.3	75.5	5.5	1.7	19%
6th	PANAMA	39.5	77.2	6.9	2.8	19%
7th	NICARAGUA	38.7	74.3	5.4	1.4	25%
8th	BANGLADESH	38.4	70.8	4.7	0.7	27%
9th	THAILAND	37.3	74.1	6.3	2.7	15%
10th	ECUADOR	37.0	75.4	6.0	2.2	22%
11th	JAMAICA	36.9	75.3	5.6	1.9	21%
12th	NORWAY	36.8	81.3	7.7	5.0	7%
13th	ALBANIA	36.8	77.3	5.5	2.2	17%
14th	URUGUAY	36.1	76.9	6.4	2.9	18%
15th	SPAIN	36.0	82.2	6.3	3.7	10%

COSTA RICA

Strong social networks, investment in health and education and a deep connection to nature may help explain why Costa Ricans are happier and live longer than the residents of most rich nations. A national commitment to environmental protection and use of renewable energy also keeps Costa Rica's Ecological Footprint small.

HPI

1st of 140

NORWAY

Universal healthcare and free education might be one reason that Norway's wellbeing and life expectancy scores are both high and some of the most equal in the world. But while Norway's Ecological Footprint is small for Europe, it's still almost three times bigger than Earth can sustain.

HPI

12th of 140

The World Happiness Report measures happiness and wellbeing on a global scale and ranks countries according to various indicators. The regions that rank the highest for happiness and wellbeing tend to be peaceful European countries with high levels of food security and access to education and healthcare. People report high levels of community trust in these locations and this is found to make them more resilient to change and challenges.

On average, people who live in cities tend to be happier than those in rural areas. This is particularly true for those living in **LEDCs**. Countries that have clear sustainable development goals are also generally happier.

Source 1

The way people rate their happiness varies depending on a range of social, economic, environmental and political factors. The Happy Planet Index is discussed further on page 85.

Top five countries for happiness according to the World Happiness Report (2017–2019)	
1	Finland
2	Denmark
3	Switzerland
4	Iceland
5	Norway

Bottom five countries for happiness according to World Happiness Report (2017–2019)	
1	Afghanistan
2	South Sudan
3	Zimbabwe
4	Rwanda
5	Central African Republic

Source 2

Global happiness top and bottom rankings

Source: Nic Marks/Happy Planet Index

Source 3

While Australia is one of the happier countries in the world, there remains a concerning disparity between the wellbeing of First Nations and non-Indigenous people.

Learning ladder G3.2

Show what you know

1 Outline what is meant by the term 'happiness'.

2 Source 1: Locate three happy and three unhappy countries on the map.

Patterns and interconnections

Step 1: I can provide short explanations for patterns and interconnections

3 Suggest why people living in urban areas may be happier than those in rural *regions*.

Step 2: I can explain patterns and interconnections

4 Source 2: Explain why some countries may rank higher than others in the World Happiness Report.

Step 3: I can use data to support explanations of patterns and interconnections

5 Source 1: List some of the least happy countries according to the map. You may need to refer to another world map for country names.

Step 4: I can use relevant sources to research further reasons for patterns and interconnections

6 Choose one of the unhappiest countries from Source 2. Research some factors that may contribute to their ranking and suggest one improvement that may help elevate their ranking.

SHEEPT, page 156
Answering questions, page 162

How is human wellbeing measured?

Given that so many factors influence a person's wellbeing, we cannot simply measure wellbeing by a country's GDP or by surveying someone's happiness. Instead, many frameworks measure wellbeing using a range of objective and subjective factors.

Objective factors are those that can be measured using numerical data such as counts or percentages. Objective measurements normally focus around health, wealth and education. For example, the Human Development Index (HDI) is an objective measurement of world regional wellbeing based on Gross National Income (GNI), average years of education and average life expectancy (as an indicator of health). On the other hand, **subjective measures** tend to be more opinion based. Communities may be asked to rank their happiness on a scale from 0–10 or may self-report mental health or life satisfaction. While objective measures may provide analysts with a more scientific and quantitative view of a country's wellbeing, subjective measures can be just as important.

Source 1

The Human Development Index (HDI) is a summary measure of key dimensions of human development: a long and healthy life, a good education, and having a decent standard of living. Scores are between 0 and 1 and higher values indicate longer life, better education and a higher standard of living.

Human Development Index, 2017

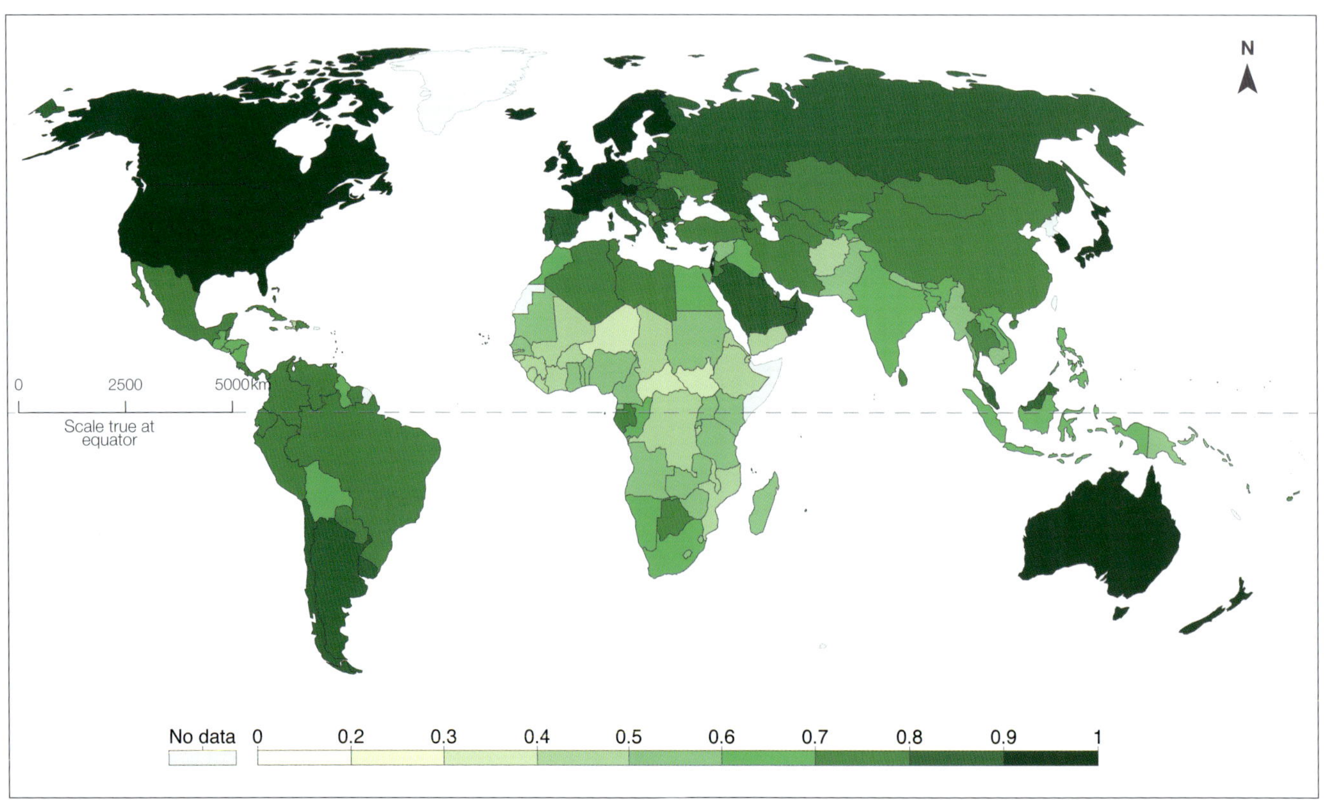

Source: Our World in Data

The Happy Planet Index (HPI) is an example of a wellbeing measurement that aims to take into account both objective and subjective factors. In a similar way to the HDI, the HPI uses objective data such as life expectancy to calculate wellbeing; however, a key difference is their focus on other contributing factors such as ecological sustainability, inequality and subjective life satisfaction ratings. As a result, these indices provide geographers with different data and distributions that can be used by policymakers to help improve their community's health and wellbeing.

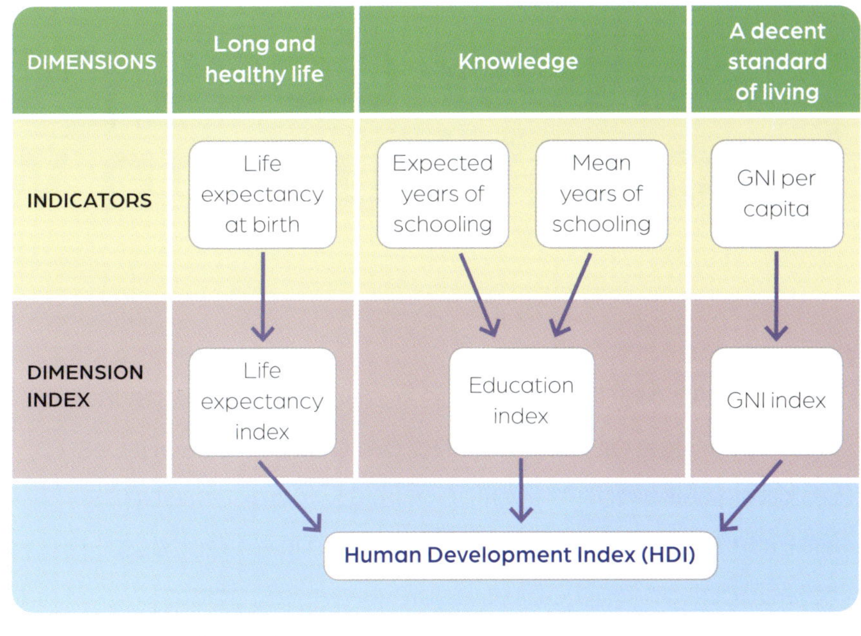

Source 2

Flow diagram indicating how the HDI is calculated

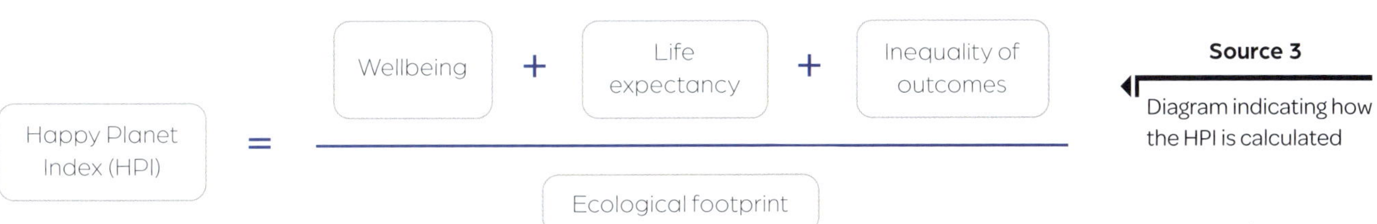

Source 3

Diagram indicating how the HPI is calculated

Learning ladder G3.3

Show what you know

1 Outline why GDP alone is not a good indicator of human wellbeing.

2 Compare HPI to HDI by constructing a table to highlight their key similarities and differences.

Digital and geospatial technologies

Step 1: I can interpret different map types using cartographic conventions

3 Access http://mea.digital/GHV10_G3_1 and click 'explore the data'. Outline how this map helps geographers understand the spatial *distribution* of happiness on a global scale.

Step 2: I can construct paper maps using correct cartographic conventions

4 Access http://mea.digital/GHV10_G3_1 again and select the table view icon at the top left. Using a blank world map and correct geographic conventions, highlight the location of the top 10 ranking countries and the bottom 10 ranking countries for HPI.

Step 3: I can access and use geospatial technology platforms such as GIS

5 Access http://mea.digital/GHV10_G3_2. Select three countries you mapped in Question 4 and complete the following table.

Country	Health	Education	Homicide rate	Carbon emissions

Step 4: I can manipulate data using digital and geospatial technologies

6 Compare the top five ranking countries for HPI shown on page 82 and compare their rankings according to HDI. Suggest any possible reasons for key similarities and differences.

Mapping with BOLTSS, page 152
PQE, page 154
Answering questions, page 162

How does poverty impact human wellbeing?

The United Nations (UN) describes **poverty** as 'more than the lack of income and productive resources to ensure sustainable livelihoods. Its manifestations include hunger and malnutrition, limited access to education and other basic services, social discrimination and exclusion, as well as the lack of participation in decision-making'. While the rates of extreme poverty have halved in the last 30 years, around the world more than 800 million people still live on less than US$1.25 per day.

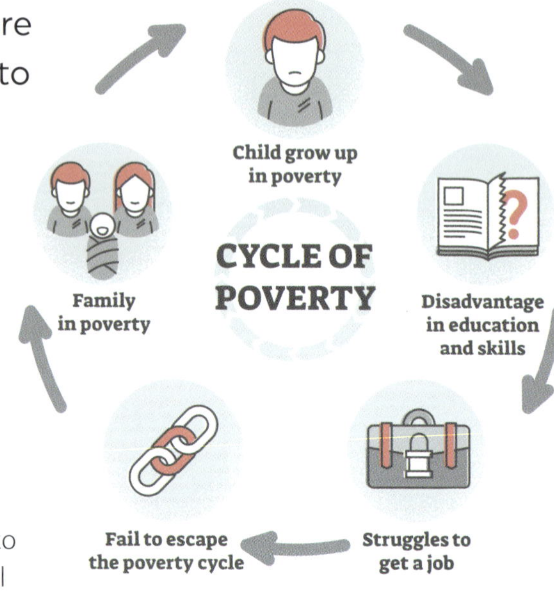

Extreme poverty rates are predicted to rise by around 88 million to 115 million people because of the COVID-19 pandemic. The global health crisis will create a 'new poor' – people living in urban areas who have lost their manufacturing or service sector jobs because of pandemic restrictions. Global warming is also going to have a huge impact on global poverty rates. In 10 years, it is estimated that our changing climate will force 68–132 million more people into extreme poverty. Currently, over 132 million people who are living in extreme poverty are located in high-flood-risk regions. As sea levels rise, these people will become climate refugees.

Source 1

Poverty is strongly interconnected with lack of access to education. More than 70 per cent of adults who live in poverty have only limited education, meaning they are more at risk of unemployment and are less able to provide for their families.

Source 2

This satellite image from 2016 shows the Earth at night. We can see the disparity in access to electricity, as some areas are far brighter than others.

Source: Alamy.com

Electricity access, 2016

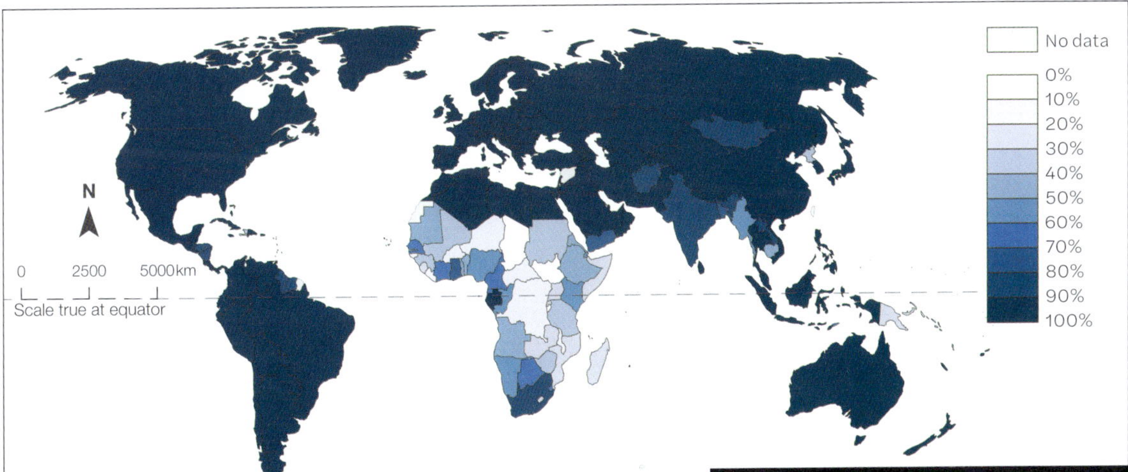

Source 3

Percentage of the population with access to electricity

Source: Our World in Data

While the term 'poverty' is usually closely connected to finances, there are other shortages that can affect people, such as a lack of electricity or 'energy poverty'. Researchers estimated that, in 2016, over 13 per cent of the world's population did not have access to electricity and more than 3 billion people did not have access to safe cooking fuels. As a result, these communities are exposed to toxic indoor pollution from alternative, unsafe fuels and have limited health care (because electricity is required to provide high-quality health care). Unsafe cooking fuel inhalation is responsible for more than four million deaths per year globally.

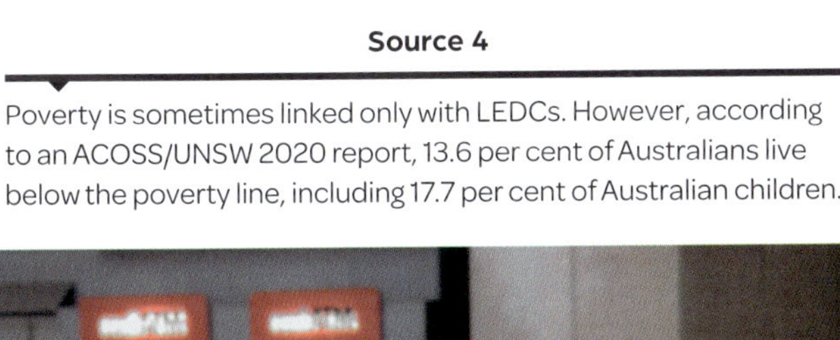

Source 4

Poverty is sometimes linked only with LEDCs. However, according to an ACOSS/UNSW 2020 report, 13.6 per cent of Australians live below the poverty line, including 17.7 per cent of Australian children.

Learning ladder G3.4

Show what you know

1 In your own words, define the term 'poverty'.

2 Summarise what is meant by the 'poverty cycle' and outline how it is interconnected with human wellbeing.

Spatial distributions and patterns

Step 1: I can identify spatial distributions and patterns

3 Source 2: Suggest how satellite imagery can help identify patterns of energy poverty on different scales.

Step 2: I can use data to quantify spatial distributions and patterns

4 Poverty is an objective measurement of wellbeing. Suggest how poverty may influence subjective ratings of happiness between people and places.

Step 3: I can describe spatial distributions and patterns

5 Locate a map of global variations in GDP from a reliable source.

a Using PQE, describe the differences in GDP on a regional *scale*.

b Describe the spatial association between high GDP and low levels of poverty in a *place*.

Step 4: I can use data to support exceptions to spatial distributions and patterns

6 'Living in poverty means more than just not having access to wealth.' Discuss this statement using data gained from reliable sources.

HOW TO

PQE, page 154
Answering questions, page 162

How is gender linked to inequality and wellbeing?

The role of women in a community plays a significant role in that community's wellbeing. In nations where maternal mortality rates, child marriages and teen pregnancies are high, and where female literacy rates, opportunities for secondary education, and labour force participation are low, levels of wellbeing are almost universally lower.

The Gender Inequality Index (GII) compares gender equality among nations through the lenses of health, empowerment and the labour market, providing insights into gender gaps around the world. In doing so, it can provide broader insights into areas that need improvement, for **MEDCs** as well as for LEDCs. For example, in Rwanda in 2016, women held 64 per cent of all available political seats in the national assembly, compared to just 19 per cent female representation in the USA. And in 2019, MEDC South Korea ranked 124 out of 149 countries in terms of labour force opportunities and participation for women, with a 35 per cent average pay difference between genders.

Gender Inequality Index (GII)

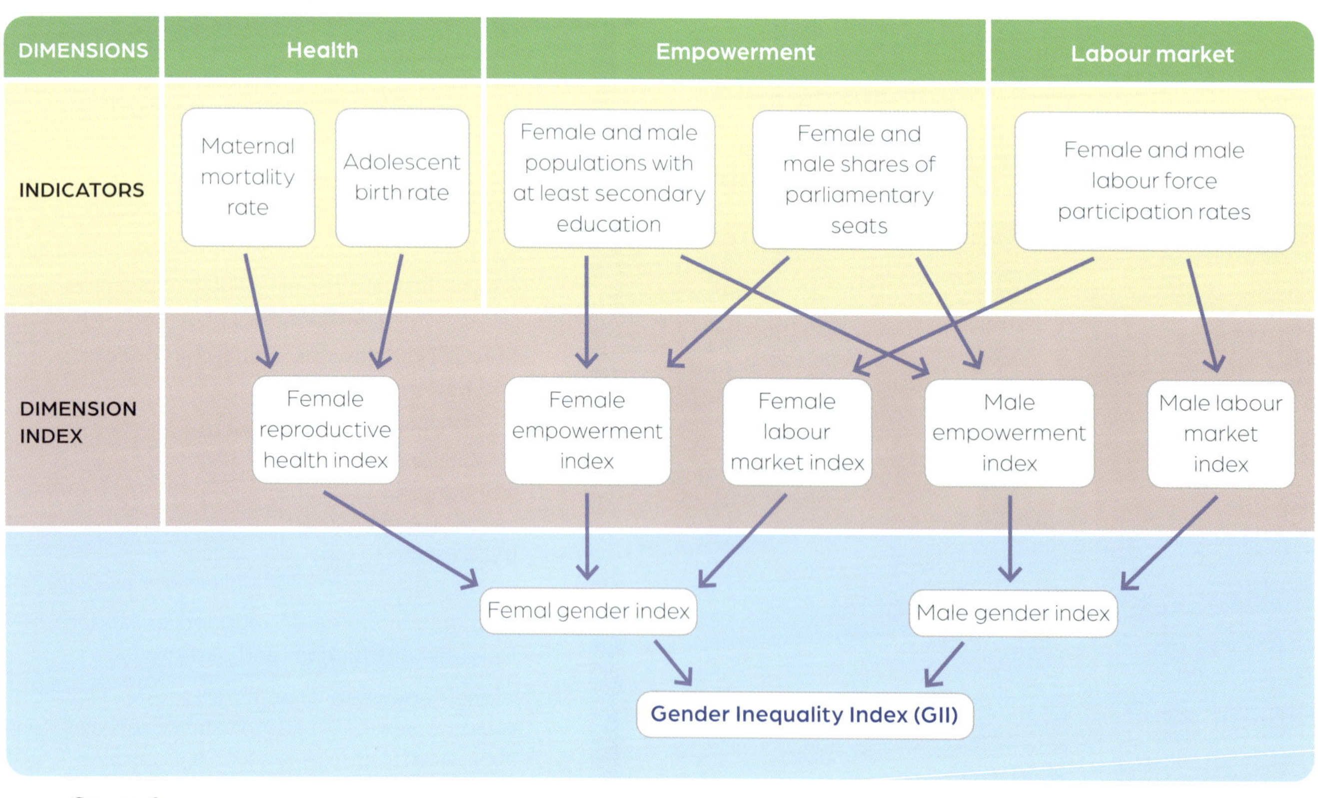

Source 1

Gender Inequality Index

2018 GII ranking	Nation	Maternal mortality ratio	Adolescent birth rate	Share of seats in parliament for women	Labour force participation rate, men	Labour force participation rate, women
1	Norway	5	5.1	41.4%	66.7%	60.2%
6	Australia	6	11.7	32.7%	70.5%	59.7%
14	New Zealand	11	19.3	38.3%	75.7%	64.6%
22	South Korea	11	1.4	17.0%	73.3%	52.8%
36	Saudi Arabia	12	7.3	19.9%	79.2%	23.4%
71	Sri Lanka	30	20.9	5.8%	72.2%	34.9%
111	Indonesia	126	47.4	19.8%	82.0%	52.2%
152	Pakistan	178	38.8	20.0%	81.5%	23.9%
189	Niger	553	186.5	17%	90.5%	67.3%

Source 2

A selection of nations and their 2018 GII ranking

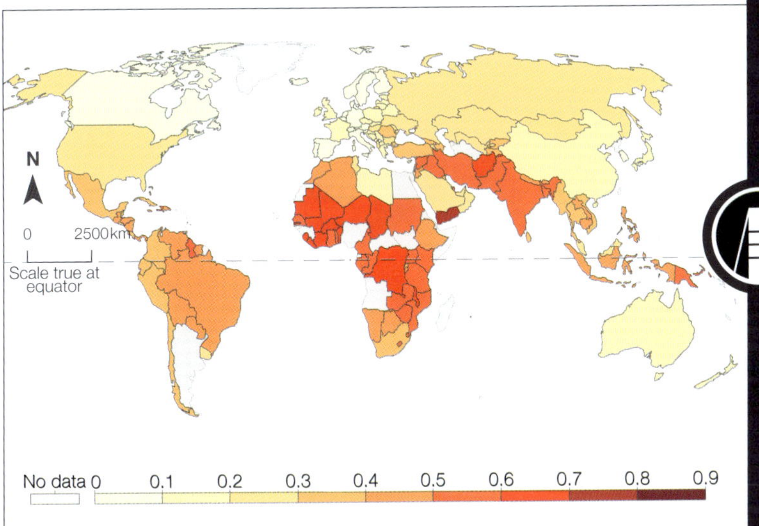

Source: Our World in Data

Source 3

Global distribution of inequality. This index covers three dimensions: reproductive health, empowerment and economic status. Scores are between 0 and 1 and higher values indicate higher inequalities.

Barriers to gender equality can vary greatly, but social and historical factors, including tradition, patriarchal power structures, religion and education, often play a major role. The impacts of inequality can include higher rates of domestic violence, lower income or social status, discrimination and more. Improving opportunities for women goes hand-in-hand with national wellbeing.

Learning ladder G3.5

Show what you know

1 In your own words, define 'gender inequality'.

2 What factors might contribute to gender inequality?

Spatial distributions and patterns

Step 1: I can identify spatial distributions and patterns

3 Source 1: Identify the five key indicators of the GII.

Step 2: I can use data to quantify spatial distributions and patterns

4 Source 2: Using the data supplied, prepare a statement to outline where low levels of gender inequality can be found.

Step 3: I can describe spatial distributions and patterns

5 Source 3: describe the global *distribution* of nations with high levels of gender inequality.

Step 4: I can use data to support exceptions to spatial distributions and patterns

6 Source 2: Niger is the lowest ranked nation in the 2018 GII, yet labour force participation rates for women are higher than for many other countries. Using the data in the table, suggest why it is outranked by other countries.

7 Saudi Arabia was ranked 36th in the world for gender equality in 2018. Using Source 3, suggest why.

HOW TO

SHEEPT, page 156
Answering questions, page 162

How is the environment interconnected with wellbeing?

Humans rely on and use the natural environment to meet our basic needs. Soil is used to plant crops, oceans are used for shipping and fishing and forests are used for natural timbers, fibres and food.

The environment provides us with the things we need to survive by maintaining its natural processes and balance. In Geography we refer to the ways that the environment helps us as '**environmental services**'. These services can be broken down into four main groups as shown in Source 1.

The future of human wellbeing is strongly interconnected with sustainable environmental use and management. As our global population continues to increase, experts predict that by 2100 we will require 80 per cent more food than we produce today. While we become more reliant on natural cycles to provide us with clean water, oxygen and suitable climatic conditions for living and farming, our current actions continue to degrade the environment.

Trade-offs are now an assumed part of policy making. While a country aims to increase its GDP and trade capacity by increasing mining and deforestation, we observe trade-offs through the decreasing ability of the environmental services to provide clean water and air, and to regulate global climate. Much of human development has come at a cost to environmental health. While these developments assist us in the short term, we rely on the productivity and services that ecosystems provide to maintain our wellbeing and health over longer time scales.

While we are protected by the structure of our society, as well as our culture and technology, we still heavily rely on the natural environment to meet our key human needs and, as a result, sustainability is crucial to our ongoing health and wellbeing.

Source 1

Humans receive a range of benefits from the natural environment.

1 **Provisioning service** Providing resources such as water, food, materials, fuels and so on

2 **Regulating service** Climate regulation, cycles of gases and resources (water cycle etc.)

3 **Cultural service** Connection to country, allowing for recreation, aesthetic value

4 **Supporting service** Photosynthesis (creating energy for producers and consumers), oxygen and soil production

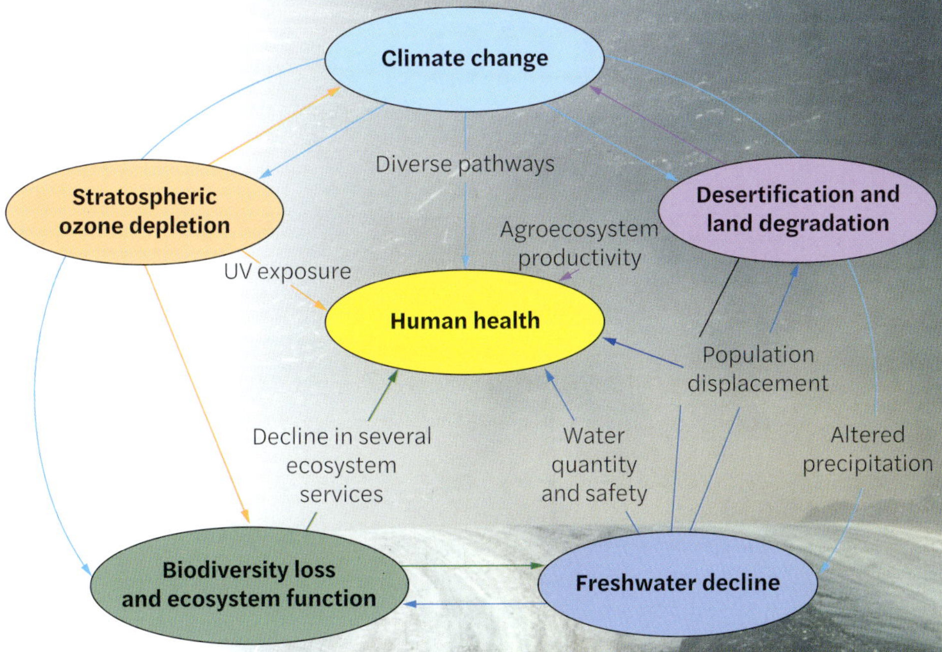

Climate change
Stratospheric ozone depletion
Diverse pathways
Desertification and land degradation
Agroecosystem productivity
UV exposure
Human health
Population displacement
Decline in several ecosystem services
Water quantity and safety
Altered precipitation
Biodiversity loss and ecosystem function
Freshwater decline

Source 2

Human health is directly affected by the environment. As this World Health Organization diagram depicts, as the different elements of the environment are depleted, they negatively affect each other and human health.

Source 3

Human actions alter environmental services.

Learning ladder G3.6

Show what you know

1 Create a photo essay that illustrates how human wellbeing is interconnected with the natural *environment*.

2 Define the term 'trade-off' with reference to economic and environmental *sustainability*.

Changes and implications

Step 1: I can identify that changes occur in the characteristics of places over time

3 Complete a field sketch highlighting key *changes* that have occurred in your local *region* over time. Annotate your sketch with the ways those changes may have impacted human wellbeing on a local *scale*.

Step 2: I can describe how places have changed over time

4 Describe how global warming may impact human wellbeing the following *scales*:
- a local
- b regional or national
- c global.

Step 3: I can explain the causes behind the change over time in a place

5 In small groups, assume the roles of the following stakeholders:
- economist
- environmentalist
- politician up for election.

Write some notes on how your stakeholder may view environmental trade-offs and their *interconnection* with human wellbeing. Compare your ideas and create a summary of some of the conflicts that we face as a community when trying to act sustainably.

Step 4: I can make predictions and outline consequences of change over time

6 Source 2: Predict how continued negative environmental *changes* can affect human wellbeing in the short and long term.

Sketches and annotating, page 158
Photo essays, page 160
Answering questions, page 162

What is the resource curse?

Human wellbeing is strongly tied to socioeconomic development. As incomes and opportunities improve, so too does access to education, gender equality and political freedoms. These developments can be fuelled by the sale of natural resources and the associated opportunities this brings. However, the reality is that the road to improved wellbeing is not necessarily straightforward.

For some nations, an abundance of natural resources does not necessarily translate to improved wellbeing; rather, it can have the opposite effect. This curious phenomenon is known as the **resource curse** (or the *paradox of plenty*). For example, the nations of Venezuela, Zimbabwe, South Africa, Nigeria, Democratic Republic of the Congo and Angola all have an abundance of different types of natural resources, including oil, gas, rare minerals, precious stones and more, yet rank low on many wellbeing and development measures. By contrast, some resource-poor nations, such as Japan, Taiwan and Singapore, are among the world's highest-ranking nations.

Source 1

Gold miners in the Democratic Republic of the Congo

Source 2

Kanpur in the state of Uttar Pradesh in India has some of the worst air quality in the world.

The reality of the resource curse is that many factors influence the use of natural resources and the potential wealth they bring. Some may be internal, including political decision making, corruption and government spending priorities – is the country's new-found wealth invested in empowering citizens with education and economic support, or is it channelled into other areas, such as the military? In some cases, money can be used to consolidate political power and influence, leading to civil conflict and instability.

External factors also play a role, including the presence of foreign investors and multilateral relationships. The very same natural resources that fuel the global economy do not necessarily translate into improved wellbeing for the nations that provide them. Furthermore, extracting natural resources can leave a legacy of environmental degradation that leads to long-term issues, such as erosion, species loss and air pollution.

On a local scale, natural resources can also put strain on communities and the environment. In Moranbah, in central Queensland, regional mining has been linked to high levels of dangerous microscopic dust particles, known as PM 2.5, and nitrogen oxides in the air, leading to long-term health risks. Nevertheless, mining remains the economic lifeblood of the community.

Similar cases can be found around the world, but are often heavily clustered in emerging economies. Of the 20 cities with the highest rates of particle matter in the air globally, 17 are found in India, Pakistan and Bangladesh. In the Indian city of Kanpur, PM 2.5 levels reach 173 micrograms per cubic metre of air. For comparison, a level of 1.5 micrograms per cubic metre of air is considered the healthy maximum. The city, located in the Indian state of Uttar Pradesh, is home to significant manufacturing and leatherworking centres.

The impacts on wellbeing and the environment in Kanpur and Moranbah are vastly different, but both are tied to the economic development of each location; the spatial variation between each is a further example of global disparity.

Learning ladder G3.7

Show what you know

1 What is the 'resource curse'?

2 Why might high levels of dust particles in the air be bad for wellbeing?

Changes and implications

Step 1: I can identify that changes occur in the characteristics of places over time

3 Using *interconnection*, suggest how an abundance of natural resources might benefit a nation.

Step 2: I can describe how places have changed over time

4 Why might the Democratic Republic of the Congo rank poorly on the Human Development Index despite having significant reserves of natural resources?

Step 3: I can explain the causes behind the change over time in a place

5 Compare Kanpur and Moranbah. Explain why both have issues with air pollution, but why the problem is significantly worse in South Asia.

Step 4: I can make predictions and outline consequences of change over time

6 Explain why a resource-poor nation, such as Singapore, might experience significantly higher levels of wellbeing and development than a resource-rich nation such as Zimbabwe.

7 Conduct further research into a country with a resource curse, such as Sierra Leone, Venezuela or Angola. Discuss the factors that seem to have prevented it from using the resource to improve wellbeing on a national *scale*.

SHEEPT, page 156
Answering questions, page 162

How does a global pandemic impact human wellbeing?

The COVID-19 pandemic has significantly altered wellbeing and happiness within and between regions. This global health crisis, coupled by an increased prevalence and intensity of extreme weather events caused by global warming, has led experts to predict that the way countries rank in future World Happiness Reports may change.

Loss of social connection, financial security and health concerns are just some of the greatest risks to wellbeing on a global scale. The 2020 World Happiness Report found that countries that had higher COVID-19-related death and illness rates, such as Italy, Spain and the USA, were generally more unhappy than those with fewer deaths. By November 2020, the USA alone had over 230 000 deaths linked to COVID-19 and were seeing more than 130 000 new cases per day.

Trust in government response and medical aid, along with stability of employment, were all factors that contributed to people's happiness and wellbeing during the pandemic. The International Labour Organization reported in April 2020 that nearly half of the global workforce was at risk of interruption, income reduction or unemployment as a result of the pandemic. Understandably, increased unemployment rates were also reported to reduce people's overall life satisfaction in 2020.

Surprisingly though, the World Happiness Report 2020 stated that some MEDCs saw an increase in happiness during the peak of the pandemic. They explained this by people's increased sense of place during the different stages of lockdown and their increased connection to community by giving and receiving support from others, as well as pride in working together towards a common goal.

Source 1

Along with other measures, masks were necessary throughout the pandemic to protect others.

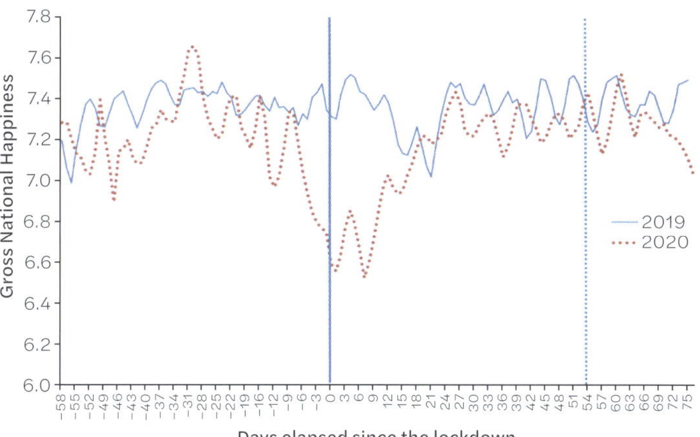

Source 2

This graph measures how Australians ranked their feelings from January to March in 2019 and 2020.

Source: Talita Greyling, Stephanie Rossouw, and Tamanna Adhikari, *South African Journal of Economics*

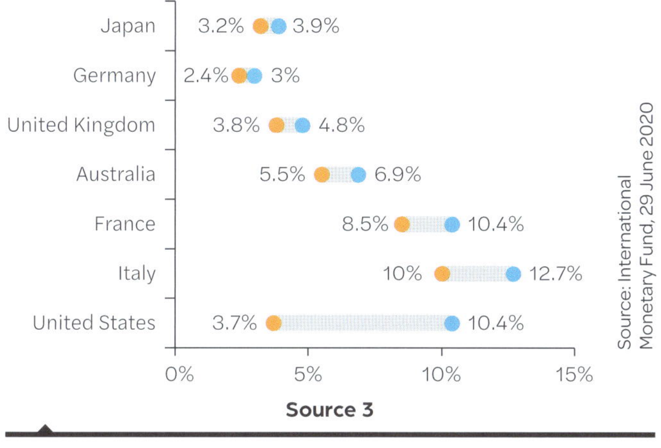

Source 3

During national lockdowns many businesses were forced to close, which led to increased unemployment across the globe.

Learning ladder G3.8

Show what you know

1 Define the term 'global pandemic'.

2 List some of the issues that the pandemic caused in Australia that may be interconnected with human wellbeing and happiness.

3 Outline how Victorians tried to create a sense of connection and belonging during the lockdowns to increase local wellbeing.

Digital and geospatial technologies

Step 1: I can interpret different map types using cartographic conventions

4 Source 2: Which map type do you think would be most suitable to show the distribution of gross national happiness during COVID-19 lockdowns on a national scale?

Step 2: I can construct paper maps using correct cartographic conventions

5 Access http://mea.digital/GHV10_G3_3. Suggest why geospatial technology is useful in analysing the spread of COVID-19 on various *scales*. Why would analysts prefer digital mapping over traditional paper maps?

Step 3: I can access and use geospatial technology platforms such as GIS

6 Access http://mea.digital/GHV10_G3_4 to view Australia's Government Stringency Index. Explain how this data gives an insight into government response to the pandemic. Reflect on and discuss with a partner how government stringency affected local wellbeing in your experience.

Step 4: I can manipulate data using digital and geospatial technologies

7 Record data in the table below by manipulating links from Questions 5 and 6.

Country	Change in unemployment	COVID-19 death rate	COVID-19 confirmed cases

HOW TO

SHEEPT, page 156
Answering questions, page 162

economics + business

Does money lead to happiness?

It is often said the 'money doesn't buy happiness'; however, some geographical data suggests that there *is* a correlation between income and self-reported life satisfaction. Stable income is required to obtain basic necessities such as food, electricity, clothing, transportation and housing. Being unemployed or without a steady income may mean a person goes without these necessities and, therefore, they experience higher stress levels and lower wellbeing.

Self-reported life satisfaction vs GDP per capita, 2017

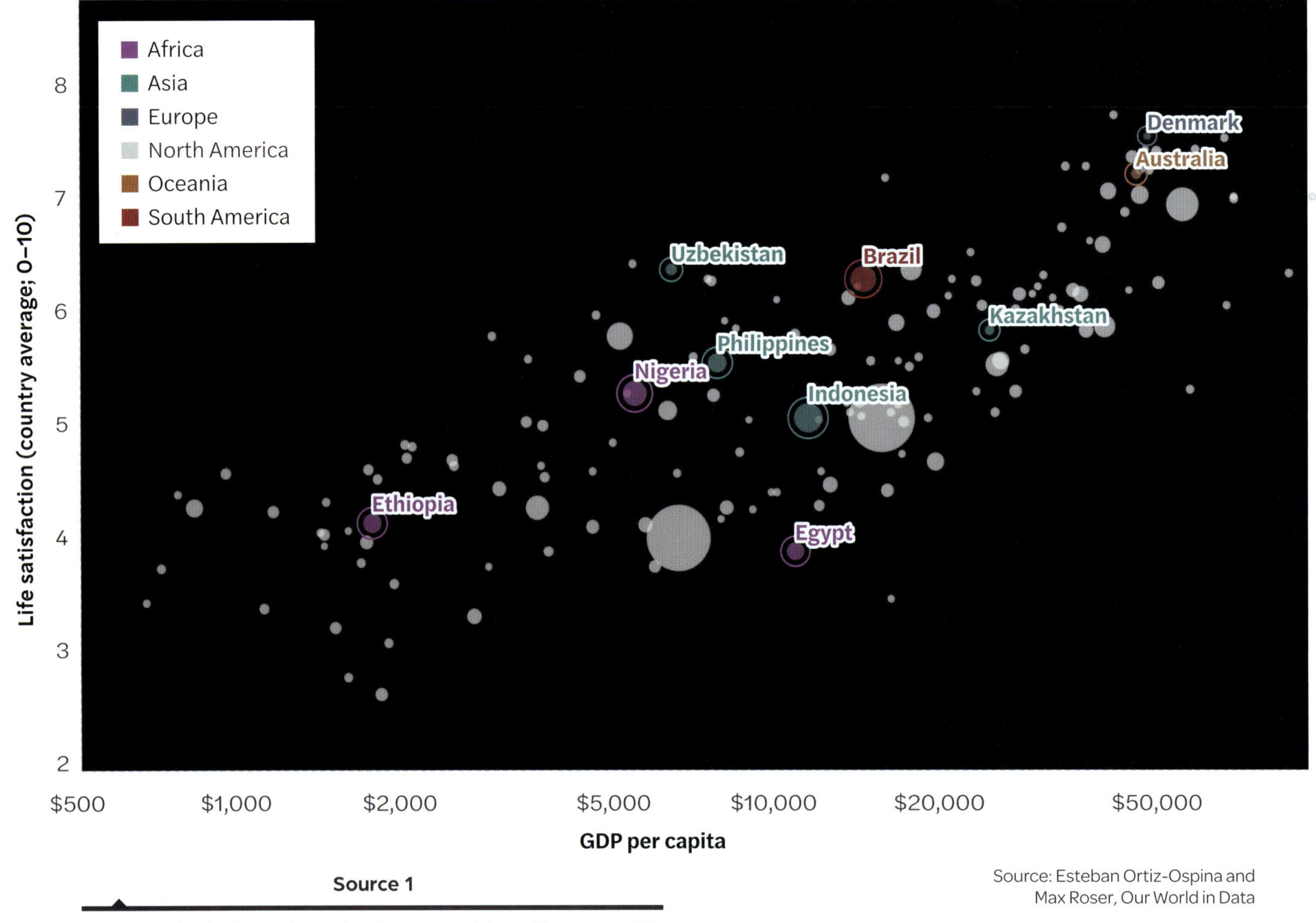

Source 1

The vertical axis shows the national average of the self-reported life satisfaction on a scale ranging from 0 –10, where 10 is the highest possible life satisfaction. The horizontal axis shows GDP per capita adjusted for inflation and cross-country price differences.

Source: Esteban Ortiz-Ospina and Max Roser, Our World in Data

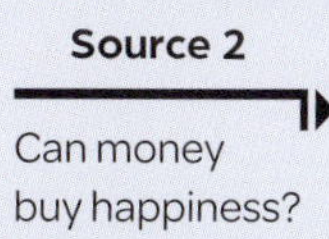

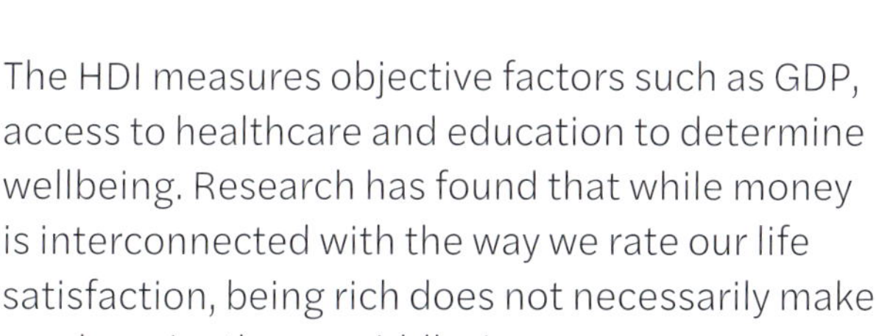

Source 2

Can money buy happiness?

The HDI measures objective factors such as GDP, access to healthcare and education to determine wellbeing. Research has found that while money is interconnected with the way we rate our life satisfaction, being rich does not necessarily make one happier than a middle-income earner.

In Australia and New Zealand, an annual income of $159 000 was found to be the limit at which more money would no longer be associated with increased satisfaction. Researchers explained that $159 000 allowed people to live comfortably, buy basic needs and pursue hobbies and travel, but excessive materialism from higher incomes was found to correlate with lower life satisfaction and lower emotional wellbeing.

In fact, some studies have found that the way we spend our money may also be linked to our wellbeing, with participants reporting higher levels of happiness when buying things for others, or donating to charity.

Source 3

Money means you are able to purchase necessary items for a comfortable life.

These broad studies highlight that, while money may be one aspect of wellbeing, other factors such as connection to people and place, living meaningful lives and caring for others also contribute to our overall happiness.

Learning ladder G3.9

Economics and business

Step 1: I can recognise economic information

1 Outline what is meant by the term 'GDP'. Research Australia's GDP for last year.

Step 2: I can describe economic issues

2 Suggest how human wellbeing is interconnected with income.

Step 3: I can explain issues in economics

3 Source 1: Choose one high-income country and one low-income country and explain the *interconnection* between their GDP and life satisfaction ranking.

Step 4: I can integrate different economic topics

4 Source 1: Consider how the data patterns may change if life satisfaction was graphed against political freedom. How would these changes in patterns help support the idea that money is only one aspect of human wellbeing?

Step 5: I can evaluate alternatives

5 'Money doesn't buy happiness.' To what extent do you agree with the statement? Write a short analytical paragraph using evidence to support your answer.

How can we improve global wellbeing?

In 2000, the UN member states signed the Millennium Development Goals (MDGs). The MDGs were a set of eight goals that aimed to end the poverty cycle and environmental degradation, to combat food insecurity and the spread of disease, and to reduce inequalities in literacy and gender by 2015.

These goals helped to lift over one billion people out of extreme poverty, reduced child mortality rates, improved global access to safe drinking water and helped reduce the spread of disease, including contributing to a 38 per cent decline in HIV infections.

But by the end of 2015 the UN acknowledged there was more to do to improve global wellbeing. As a result, the 2030 Agenda for Sustainable Development Goals (SDGs) were established. It comprised 17 goals to encourage nations to continue to work towards ending inequality and poverty, while also focusing on environmental sustainability and climate change. The SDGs were developed in consultation with a range of stakeholders to ensure they are relevant to people and place. The SDGs have a stronger focus on human rights and inequalities on a global scale with explicit reference to vulnerable groups of people.

Source 1

Sustainable development goals developed by the UN

2 ZERO HUNGER

13 CLIMATE ACTION

Challenges of making an equitable world

The goals established present a range of challenges on a global scale. For example, experts estimate that eradicating poverty may cost around $66 billion a year and creating sustainable and useable infrastructure may cost over $7 trillion globally. Maintaining peace will mean ending decades-long conflicts and finding homes for millions of displaced people. Within the 17 overarching goals there are 169 targets that the SDGs aim to achieve. Further, some of these are not quantifiable and so it will be hard to measure change over time.

The third goal is to ensure good health and wellbeing on a global scale. However, the global COVID-19 pandemic has halted and even reversed some of the work to meet this target by limiting access to and effectiveness of health services and interrupting key vaccination programs. To try to protect health services, many countries implemented strict lockdowns. This resulted in an estimated 71 million people being pushed into extreme poverty, while also putting additional pressure on trade, food and resource production.

Source 2

Infographic showing the impact of the pandemic on SDG 3

Learning ladder G3.10

Show what you know

1 Outline some key differences between the MDGs and SDGs.

2 Source 1: Summarise the 17 SDGs and how they will help improve human wellbeing on a global *scale*.

Patterns and interconnections

Step 1: I can provide short explanations for patterns and interconnections

3 Go to the SDGs tracker at http://mea.digital/GHV10_G3_7. List three countries that are less than 50 per cent of the way to achieving the SDGs. Provide two general reasons as to why that may be the case.

Step 2: I can explain patterns and interconnections

4 Access http://mea.digital/GHV10_G3_6. Using SHEEPT, explain why some countries are further away from eradicating poverty by 2030 that others.

Step 3: I can use data to support explanations of patterns and interconnections

5 Access http://mea.digital/GHV10_G3_5. Explain the spatial *distribution* of subjective wellbeing using two SHEEPT factors.

Step 4: I can use relevant sources to research further reasons for patterns and interconnections

6 Using reliable resources, research other ways global crises such as the COVID-19 pandemic or global warming may impact our ability to reach the SDGs by 2030.

SHEEPT, page 156
Answering questions, page 162

G3.11

civics + citizenship

What is meant by 'the common good'?

'The common good' is the state in which the common interests of all people are met as a result of equal opportunities, benefits and access to facilities. Acting for the common good can guide our moral compass to act in the best interests of the community.

There are three main elements that make up the common good:

1 rights: respect for others and willingness to help others
2 needs: ability to meet basic human needs and help to access these resources where required
3 peace: a feeling of safety, security and connectedness with others.

A state of common good is not easy to achieve. It assumes we have social connections and can depend on each other. It also assumes that we are willing to make sacrifices for others and are willing to treat each other with goodwill. Prejudice, ego and self-interest can complicate the common good, as these tend to encourage selfish behaviours and actions.

You may consider the common good in everyday decisions without even realising it. For example, during the 2020 COVID-19 Stage 4 lockdowns, individuals needed to sacrifice their own livelihoods and social connections to isolate, socially distance and get tested, even with mild symptoms. In order to achieve widespread common good, we rely on leadership and decision makers to create policies that benefit people within and between regions.

As we work towards vaccinating people against COVID-19, another moral dilemma arises. If the vaccine is only distributed to a few wealthy nations, few will benefit. In order to avoid another global outbreak, the vaccine needs to be made available to all people no matter where they live. In October 2020, 115 international personalities and 19 Nobel Laureates requested that the vaccine be made available free of charge to ensure that there is an equitable distribution of treatments to ensure a global 'common good'.

Source 1

Love Your Neighbour is a socially conscious clothing label from Switzerland. Its products are certified or accredited by a number of schemes that ensure ethical and environmental practices and 12 per cent of their profits are donated to homeless charities.

Source 2

During COVID-19 lockdowns we all made sacrifices to ensure the common good by quarantining to protect the health of the community. Many families remained apart for months to ensure the safety of others.

Source 3

It's not always easy to make a decision for the common good.

Learning ladder G3.11

Civics and citizenship

Step 1: I can identify topics about society

1 Summarise what is meant by 'the common good' and how it can be achieved.

Step 2: I can describe societal issues

2 Outline some key issues when trying to achieve a 'common good' on a global *scale*.

Step 3: I can explain issues in society

3 Suggest how the sustainable development goals (pages 98–9) are working toward the common good on a global *scale*.

Step 4: I can explain different points of view

4 Consider how political systems may impact the ability to achieve a common good in different world *regions*.

Step 5: I can analyse issues in society

5 Provide another example of how you consider the common good in your everyday actions. What are some conflicts that may inhibit you from always working towards the benefit of others?

How does wellbeing vary on a local scale?

It is easy to envision wellbeing through an international lens, dividing the world into more and less developed countries, between the haves and have nots. The reality is that disparity in wellbeing is common in virtually all contexts and scales, be it in terms of income, access to services, gender equality, personal freedoms, education levels, emotional wellbeing and more.

In Melbourne, housing is a key example of wellbeing and inequality. In 2020, the Victorian Public Tenants Association estimated that up to 100 000 people were on the waiting list for public housing, including those from low socioeconomic backgrounds, new migrants and those fleeing domestic violence. In central Melbourne, public housing towers dot the skyline in inner suburbs such as Flemington, South Melbourne and Fitzroy. For decades, they offered space for thousands of occupants, many having waited long periods for accommodation.

In early July 2020, Victorian Premier Daniel Andrews ordered nine such towers in Flemington and North Melbourne into immediate and total lockdown after a surge in reported positive cases of COVID-19.

Source 2

Residents in Malvern enjoying public spaces during lockdown

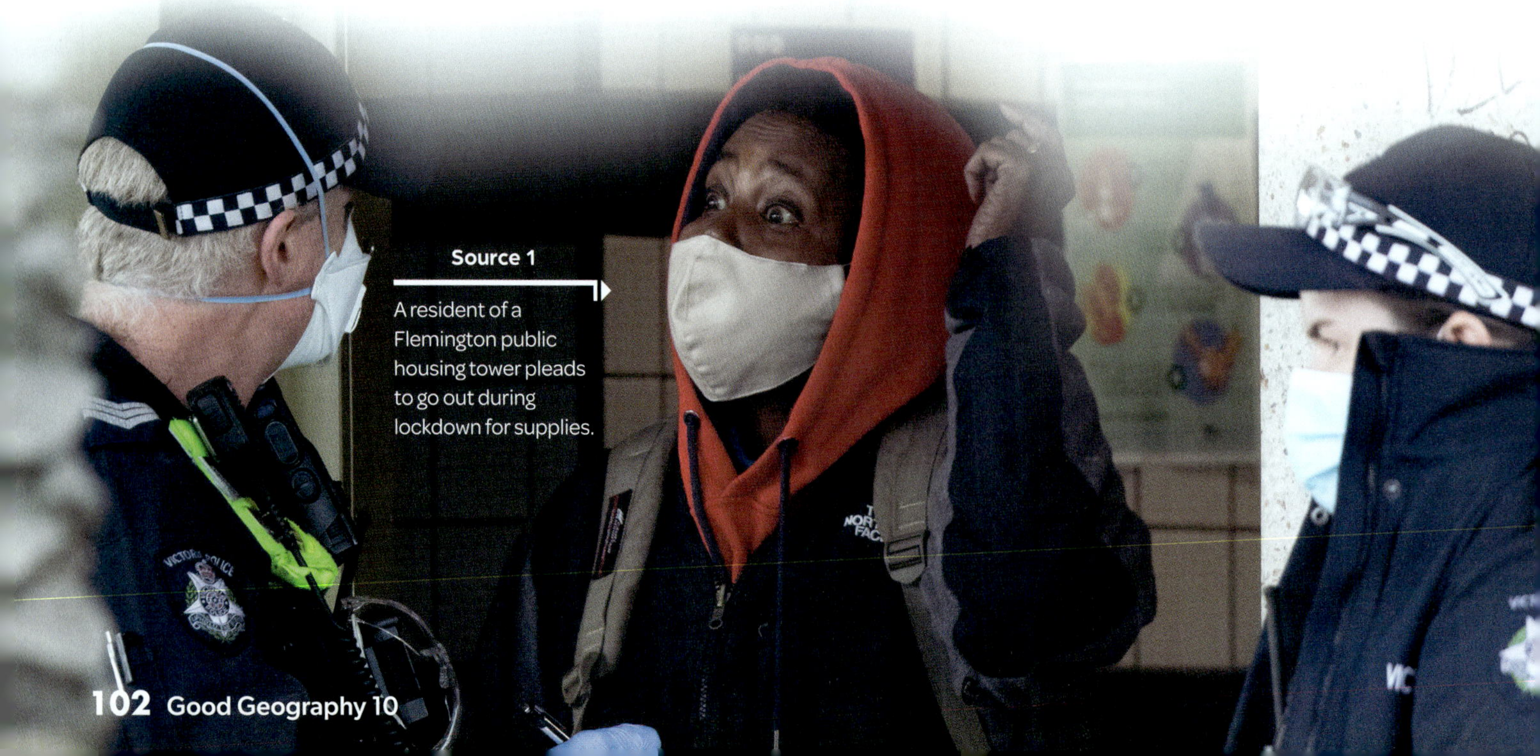

Source 1

A resident of a Flemington public housing tower pleads to go out during lockdown for supplies.

Authorities argued that the towers contained conditions similar to cruise ships: confined spaces with large numbers of people and small shared common areas – ideal for spreading the contagious virus. For days, residents remained under strict lockdown in cramped conditions and the resulting media coverage revealed significant challenges for residents, notably ongoing wellbeing concerns and the difficulties for authorities in adequately supporting people's diverse linguistic and cultural needs.

By contrast, as Stage 4 lockdowns followed for the Greater Melbourne region, residents in other suburbs were required to remain within a given radius of their home. For many areas, this still afforded plentiful opportunities for exercise and access to most services. In the affluent inner southeast, one resident complained that she had already 'done all of Brighton' on her daily walks, while others complained of having become tired of Netflix.

The disparity of the lockdown experience between those in public housing and those in their own homes with access to public spaces is a strong example of how wellbeing can vary on a local scale. According to the Black Dog Institute, there is a clear link between access to green spaces and mental health; limited access can lead to increased stress and anxiety. Across Melbourne, access to green spaces varies significantly, as seen in Source 3.

Access to parkland within permitted 5 kilometre radius, Melbourne 2020

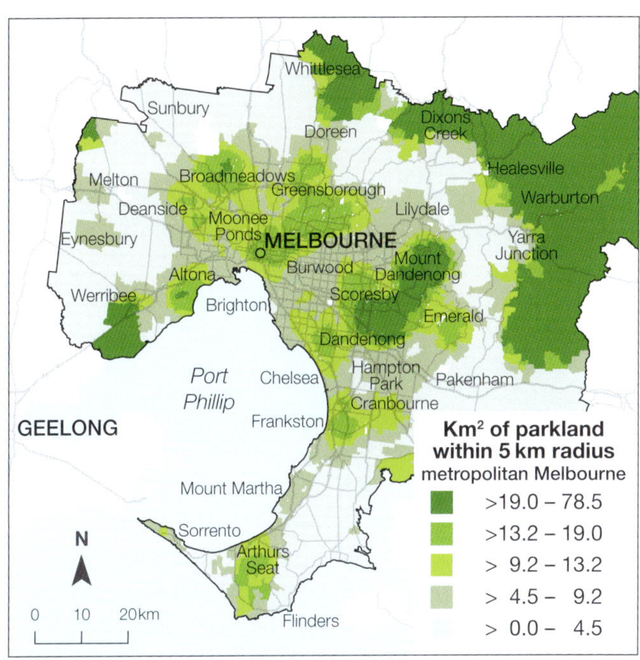

Source 3

Grey and light green spaces indicate limited access to parklands.

Source: Matilda Education Australia/Custom Mapping Services. Data from Ali Lakhani and Dennis Wollersheim.

Learning ladder G3.12

Show what you know

1 Why did the Victorian Premier order the public housing towers in Flemington into lockdown in 2020?

2 How would the lockdown experience have been different for those in free-standing homes and those in smaller accommodation? How might the lockdown experience have contributed to stress and anxiety?

Patterns and interconnections

Step 1: I can provide short explanations for patterns and interconnections

3 Why do you think Melbourne's public housing towers are located in the inner suburbs, rather than the outer suburbs?

Step 2: I can explain patterns and interconnections

4 The authorities referred to the towers as 'cruise ships in the sky'. Explain how their design increases the risk of spreading viruses.

Step 3: I can use data to support explanations of patterns and interconnections

5 Describe the *distribution* of areas with limited green space in Source 3. Suggest how this pattern might contribute to inequality of wellbeing.

Step 4: I can use relevant sources to research further reasons for patterns and interconnections

6 According to Source 3, the availability of parkland across Melbourne varies significantly. Collingwood, for example, has just 0.4 square metres of parkland per person and Fitzroy just 2 square metres, while other suburbs have far more. Explain how this might have affected wellbeing for residents in different parts of Melbourne during the lockdown, where people had to stay within 5 kilometres of their home.

PQE, page 154
SHEEPT, page 156
Answering questions, page 162

G3.13

civics + citizenship

How can we address inequality on a local scale?

Across Australia, on any given night, nearly 100 000 people are homeless. Almost a quarter of these can be found in urban Victoria, primarily in Melbourne.

Homelessness remains an ongoing issue in Melbourne, one that was thrown sharply into focus during the 2020 COVID-19 lockdowns, when 2000 homeless people were provided with emergency accommodation. When the pandemic was announced, the Victorian Government allocated $6 million of crisis funding for agencies that procure temporary housing. Furthermore, $150 million was budgeted for long-term shelter – including hotel accommodation – for rough sleepers through a Victorian Government scheme known as 'Home for the Homeless'.

Ongoing accommodation has proved to be a boon to the wellbeing of those involved, who have reported reduced rates of anxiety, depression and other ongoing health concerns. Furthermore, stable shelter has provided a platform for individuals to overcome addictions, seek stable employment and re-establish routines and connections with family and friends. Long-term funding, however, remains an ongoing concern, with the State Government reviewing the program's structure and exploring a range of accommodation options, rather than continuing to fund expensive hotel rooms.

Homelessness is traditionally associated with people sleeping rough on the streets, yet this represents only a fraction of the actual issue. As seen in Source 2, just 7 per cent of the state's homeless in 2016 were sleeping rough. Homelessness can actually be divided into three categories:

1 Primary: People without conventional housing. This includes people living rough on the streets, squatting illegally, sleeping in improvised shelter or in their car.

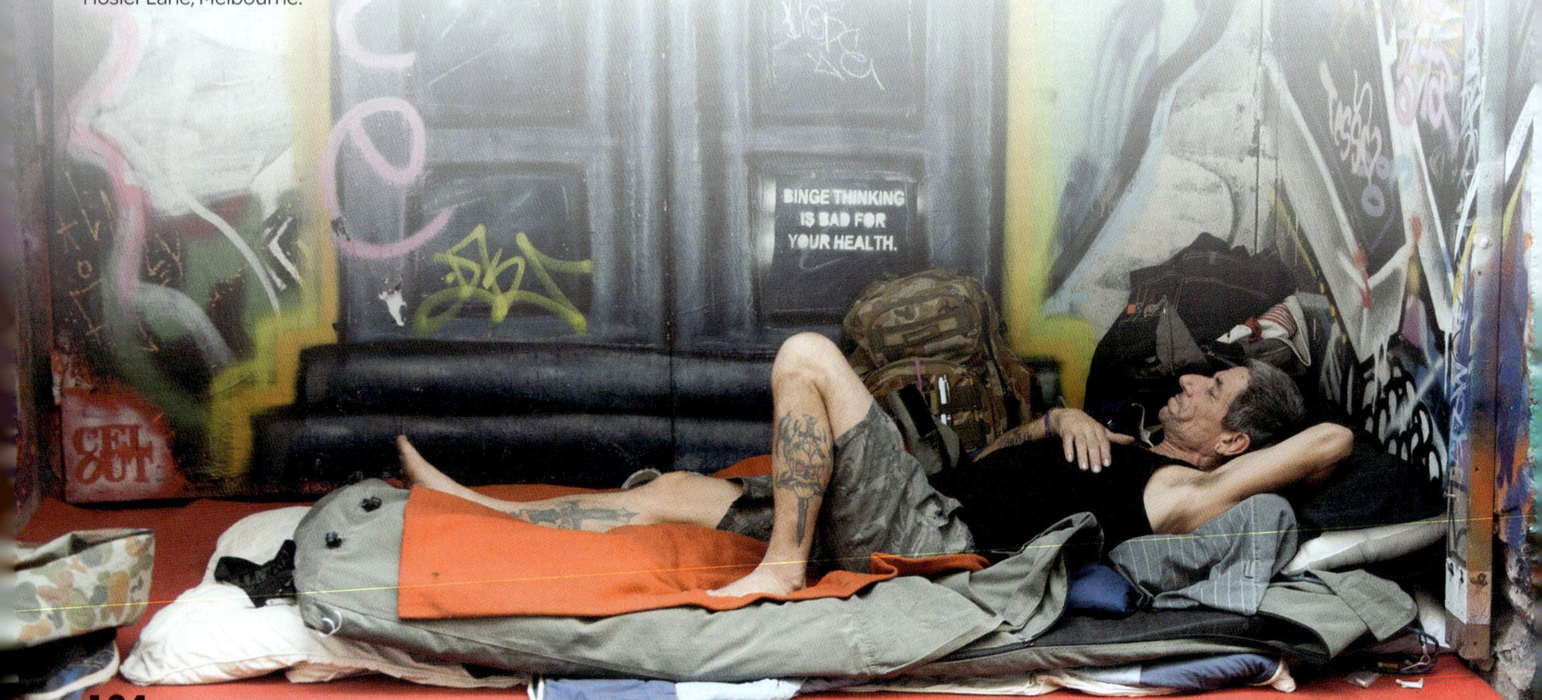

Source 1

A homeless man rests in Hosier Lane, Melbourne.

Persons in improvised dwellings, tents or sleeping out	1118	7%
Person in supported accommodation for the homeless	7175	45%
Person staying temporarily with other households	3078	19%
Persons living in boarding houses	4411	28%
Persons in other temporary lodgings	108	1%
Total	15 890	100%

Source 2

Number of homeless, Victoria (revised), Census data. The 2016 Census put the number of homeless in Victoria at almost 16 000 but the number is certainly higher because of the difficulties in sourcing accurate data.

2 Secondary: People who move frequently between accommodation, staying with friends or family, or staying in crisis housing or short-term boarding houses.

3 Tertiary: People living in mid- to long-term boarding accommodation or caravan parks without the security of a lease or private facilities.

Addressing the factors behind homelessness remains key to addressing inequality on a local scale. Factors that lead to homelessness may include:

- marital/family breakdowns
- domestic violence
- alcohol/substance abuse
- mental health issues
- unemployment
- eviction/loss of accommodation
- past or recurring trauma.

Across the state of Victoria, there are a broad range of government services, programs and initiatives to try to address these issues, as well as a host of non-government organisations, charities and religious missions. One such provider is the City of Melbourne, which facilitates the following:

- a dedicated homelessness support team to provide welfare
- weekly meetings with other organisations to support complex cases
- connecting with grassroots organisations to help the homeless
- providing free training for local businesses to better support and understand the homeless community
- funding appropriate organisations that support the homeless
- preparing a guide for how to find and access budget services
- conducting a biannual *Streetcount* event to gather demographic data on those sleeping rough across Melbourne.

Learning ladder G3.13

Show what you know

1 How many people were homeless in Victoria, according to the 2016 Census?

2 How does the City of Melbourne support the homeless?

Civics and citizenship

Step 1: I can identify topics about society

3 In your own words, define 'homelessness'.

Step 2: I can describe societal issues

4 Describe the three different classifications of homelessness.

Step 3: I can explain issues in society

5 Working with a partner, select one of the factors that can lead to homelessness. Prepare a flow chart outlining how.

Step 4: I can explain different points of view

6 Some believe that State Government funding shouldn't be used to pay for hotel accommodation for the homeless. Prepare a brief letter to the editor of your local newspaper to outline your stance.

Step 5: I can analyse issues in society

7 As house prices increase and our population grows, homelessness is growing in Melbourne. Create a mind map to demonstrate why this can create further issues and why inequality is increasing.

G3.14

getting involved

What is *The Big Issue*?

Addressing inequality on a local scale can be considered grassroots work. It involves engaging with the local community on an individual level with an understanding of local sociocultural contexts and building effective long-term relationships. In Melbourne, *The Big Issue* is synonymous with support and empowerment, providing opportunities for positive steps forward.

Based on a similar concept in the UK, the magazine *The Big Issue* launched on the steps of Flinders Street Station in 1996. Essentially, disadvantaged people buy copies of the magazine for $4.50 and then sell them for $9, keeping the difference. The Big Issue organisation then reinvests profits into social programs that support the individuals involved and the broader community. The magazine itself has a wide range of articles, including social justice topics and vendor profiles, and has sold more than 13 million copies since arriving in Australia, providing $30 million worth of income for disadvantaged Australians.

In addition to providing a source of income for individuals who may experience limited opportunities in the traditional workplace, the magazine provides empowerment – a chance for vendors to improve wellbeing in a number of ways.

In terms of social enterprises, The Big Issue organisation also runs:

- The Women's Subscription Enterprise: work, training and opportunities for homeless, disadvantaged and marginalised women
- The Community Street Soccer Program: Weekly soccer training and support at 16 locations across Melbourne, including in two correctional facilities, that provides opportunities for social inclusion for homeless, disadvantaged and marginalised men and women
- The Big Issue Classroom: An outreach program into schools for students to learn from speakers about the experiences of being homeless and disadvantaged
- The Big Idea: A social enterprise planning competition for universities to come up with effective strategies to support homeless, disadvantaged and marginalised people
- Homes for Homes: An initiative to support affordable housing through encouraging donations from real estate sales to fund social housing projects.

In addition to the more than $30 million in income for vendors, *The Big Issue* and its social enterprise programs have had a tremendous impact on the homeless, disadvantaged and the marginalised in Melbourne, including:

- demand for social, health and justice services from vendors has dropped significantly, saving the community approximately $25 000 per vendor
- the Street Soccer program has had a positive impact on the physical and mental health of participants, decreasing demand for welfare support, hospitals and correctional facilities, reducing societal costs by up to $7 million dollars annually
- more than 160 women have achieved employment through the Women's Subscription Enterprise across Australia
- more than 4500 guest speaker opportunities have been provided through the Big Issue Classroom, reaching more than 100 000 students.

Source 1

A *The Big Issue* vendor in Melbourne

Getting involved

1 Create a mind map to show the benefits for vendors of selling *The Big Issue*.

2 Using data, suggest how society can broadly benefit by reducing inequality.

3 Prepare a SWOT analysis looking at the success of *The Big Issue*. Based on your chart, to what extent do you feel it can be considered a success?

4 Complete an audit to explore how your school supports social enterprise.
   a What charities or organisations does your school support?
   b What fundraising opportunities are undertaken throughout the year (perhaps through free-dress days or sports carnivals)?

5 a In groups of two or three, brainstorm viable opportunities for further fundraising or social enterprise. How might the school be able to support marginalised individuals or groups?
   b Choose one of your ideas from Question 5a. Have each group produce an 'elevator pitch' – with 30 seconds to explain the idea to the class.
   c As a class, vote on the best ideas (more than one is possible).
   d Based on your class results, draft a brief proposal for your principal – how might the school be able to better support marginalised or disadvantaged groups? Ensure you highlight the benefits for individuals and society.
   e Prepare a brief reflection – why is social enterprise and empowerment important?

For more information on *The Big Issue* access http://mea.digital/GHV10_G3_8.

Masterclass

Learning ladder

Work at the level that is right for you or level-up for a learning challenge!

Self-reported life satisfaction, 2018

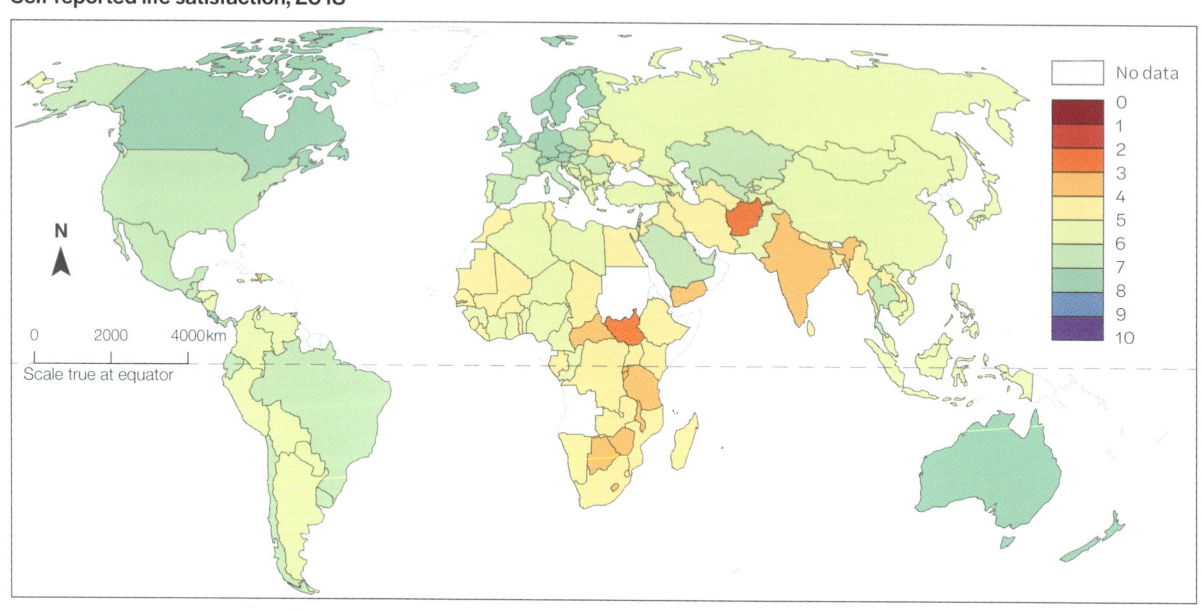

Source 1

Self-reported life satisfaction according to the World Happiness Report, with 0 the lowest and 10 the happiest.

Source: Esteban Ortiz-Ospina and Max Roser, Our World in Data

Share of population living in extreme poverty by world region

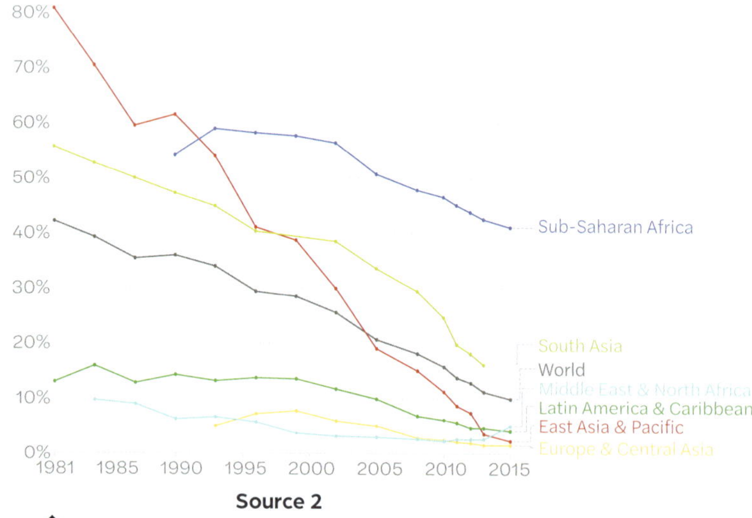

Source 2

Extreme poverty is defined as living with less than $1.90 per day in 2011 international dollars. International dollars are adjusted for price differences across countries and time.

Source: Esteban Ortiz-Ospina and Max Roser, Our World in Data

Source 3

Cars lining up so people can be tested for COVID-19 in Florida, USA

Step 1

a I can identify spatial distributions and patterns

Source 2: Describe the overall trend for the share of people living in extreme poverty on a world *scale*.

b I can provide short explanations for patterns and interconnections

Source 1: Using two SHEEPT factors, explain why life satisfaction varies globally.

c I can identify that changes occur in the characteristics of places over time

Source 3: The USA is considered an MEDC. Identify two ways that COVID-19 has put pressure on existing infrastructure and systems.

d I can list primary and secondary methods useful for my study

List three secondary sources that provide reliable objective human wellbeing data.

e I can interpret different map types using cartographic conventions

Explain why a choropleth map is the most suitable for the data displayed in Source 1.

Step 2

a I can use data to quantify spatial distributions and patterns

Source 1: Using PQE, describe the *distribution* of life satisfaction worldwide.

b I can explain patterns and interconnections

Source 2: Using three SHEEPT factors, explain why the share of people living in extreme poverty has decreased in East Asia and the Pacific *regions*.

c I can describe how places have changed over time

Summarise how a focus on 'the common good' may change a place over time.

d I can successfully use data collection methods

Create a data bank that quantifies how the world is currently working towards the sustainable development goals.

e I can construct paper maps using correct cartographic conventions

Using a suitable legend and a blank world map, illustrate the share of people living in extreme poverty in 2015 on a global *scale*.

Step 3

a I can describe spatial distributions and patterns

Source 3: Describe the *distribution* of cars.

b I can use data to support explanations of patterns and interconnections

Create a photo essay that illustrates key factors necessary for human wellbeing and happiness.

c I can explain the causes behind the change over time in a place

Explain how COVID-19 has impacted more than just human health on a global *scale*.

d I can filter collected data

Source 2: Suggest why the person that drew this graph grouped data according to world *region* and not by country.

e I can access and use geospatial technology platforms such as GIS

Locate an interactive map that allows you to explore the spread of COVID-19 on various *scales*. Outline how this data is useful for governments and policymakers.

Masterclass

Step 4

a I can use data to support exceptions to spatial distributions and patterns

Source 2: Using PQE, compare the pattern of people living in extreme poverty over time for two world *regions*.

b I can use relevant sources to research further reasons for patterns and interconnections

Chose a country that interests you. Using reliable resources, create a poster that explores both objective and subjective data for human wellbeing in this *place*.

c I can make predictions and outline consequences of change over time

Predict how human wellbeing may *change* over time as a result of global warming.

d I can organise data collected according to relevance for a research question

Complete a field sketch that illustrates key infrastructure or zones in your local *region* that are important for human wellbeing and happiness.

e I can manipulate data using digital and geospatial technologies

Access http://mea.digital/GHV10_G3_9. Is this objective or subjective data? How is this data useful in inferring human wellbeing?

Step 5

a I can identify multiple spatial distributions and patterns

Source 1, page 96: Identify the *interconnection* between GDP and life satisfaction in different world *regions*.

b I can interpret causes of patterns and interconnections

Explain any *interconnections* highlighted in Step 5, Question a.

c I can interpret data to quantify predictions based on research

Source 3: Consider the SHEEPT factors that could apply to this image. Predict how the COVID-19 pandemic may alter the USA in the future.

d I can evaluate the success of research methods

Evaluate the effectiveness of the HPI in calculating human wellbeing on a global *scale*.

e I can draw conclusions from geographical information in digital and geospatial technologies

Explore data for the varying measures of human wellbeing such as the HPI and HDI. Based on the data presented, do you agree with the World Happiness Report (2019) that Finland is the happiest place on Earth? Justify your answer.

Capstone

How can I understand development and disparity?

In this chapter, you have learned a lot about development and disparity. Now you can put your new knowledge and understanding together for the capstone project to show what you know and what you think.

In the world of building, a capstone is an element that tops off a building or a wall. That is what the capstone project will offer you, too: a chance to top off and bring together your learning in interesting, critical and creative ways. You can complete this project yourself, or your teacher can make it a class task or a homework task.

mea.digital/GHV10_G3

Scan this QR code to find the capstone project online.

Place studies of human wellbeing

G4

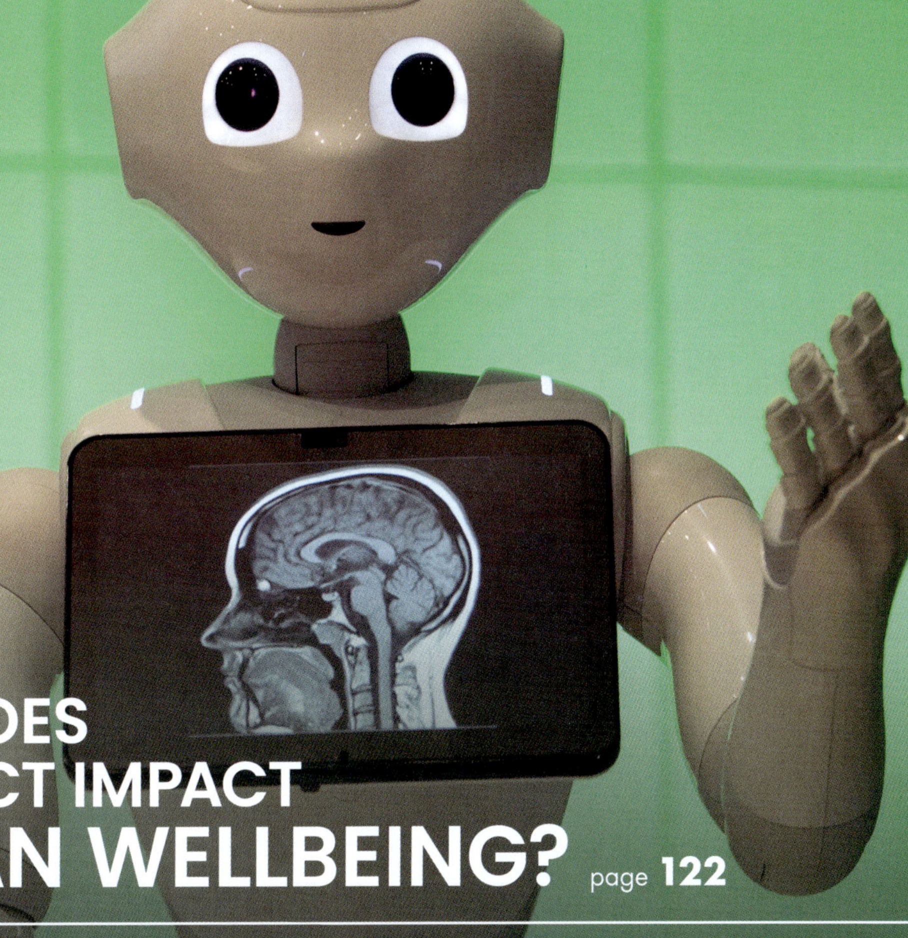

HOW DOES CONFLICT IMPACT HUMAN WELLBEING?

page 122

spatial distributions and patterns

page 114

WHY ARE NORDIC PEOPLE SO HAPPY?

changes and implications

page 118

HOW IS JAPAN IMPROVING COMMUNITY WELLBEING?

patterns and interconnections

page 132

HOW DOES HUMAN WELLBEING VARY BETWEEN COMMUNITIES IN SOUTHERN AFRICA?

How can I learn about human wellbeing using place studies?

To understand human wellbeing, we need to explore the factors that influence people's access to key resources and their happiness, on a variety of scales. Place studies give us an insight into these factors by allowing geographers to compare and contrast different world regions based on objective and subjective data.

learning ladder

Step	Spatial distributions and patterns	Patterns and interconnections	Changes and implications
step 5	**I can identify multiple spatial distributions and patterns** I can take my PQE one step further to find links or relationships that exist in place studies of human wellbeing.	**I can interpret causes of patterns and interconnections** I can use multiple sources to find links or relationships that exist in place studies of human wellbeing and can explain 'Why?'.	**I can interpret data to quantify predictions based on research** I can use external data from research as evidence of the positive and negative impacts of a change I have predicted.
step 4	**I can use data to support exceptions to spatial distributions and patterns** I can use data to answer 'Why?' about the exceptions identified in a PQE analysis of place studies.	**I can use relevant sources to research further reasons for patterns and interconnections** I can use sources other than this textbook to further research patterns I observe in place studies.	**I can make predictions and outline consequences of change over time** I can use my knowledge of natural processes and world regions to make an educated guess about the positive and negative impacts of change in place studies.
step 3	**I can describe spatial distributions and patterns** I can describe patterns, quantify them and point out exceptions (PQE) to describe place studies.	**I can use data to support explanations of patterns and interconnections** I can use data from a map or graph to explain patterns I observe in place studies.	**I can explain the causes behind the change over time in a place** I can use my knowledge of natural processes and world regions to explain why changes may occur over time in place studies.
step 2	**I can use data to quantify spatial distributions and patterns** I can read data and use it to measure key trends on a map or graph about place studies.	**I can explain patterns and interconnections** I can identify social, historical, economic, environmental, political and technological (SHEEPT) factors to help me explain place studies.	**I can describe how places have changed over time** I can use specific examples to describe changes over time in place studies.
step 1	**I can identify spatial distributions and patterns** I can find key trends on a map or graph about place studies.	**I can provide short explanations for patterns and interconnections** I can write descriptions of patterns and interconnections that I find in place studies.	**I can identify that changes occur in the characteristics of places over time** I can read information and answer questions about changes over time in place studies.

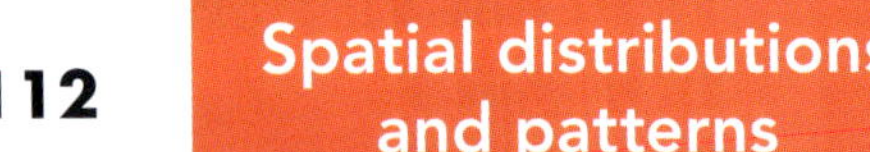

Source 1

Differences in living conditions in Johannesburg, South Africa

I can evaluate the success of research methods
I can look back and comment on the data collection methods I used, and evaluate how successful they were in helping me answer a research question about place studies.

I can organise data collected according to relevance for a research question
I can review the data I have collected in the field and display it using graphs, tables, annotations and captions.

I can filter collected data
I can review my collected data and select the most relevant data to answer a research question about place studies.

I can successfully use data collection methods
I can use primary and secondary data collection methods in the field and classroom to investigate place studies.

I can list primary and secondary methods useful for my study
I can create a checklist of methods to investigate place studies and categorise them as primary or secondary methods.

Communicate data

I can draw conclusions from geographical information in digital and geospatial technologies
I can interpret and analyse patterns by using different layers and features on geospatial technology platforms.

I can manipulate data using digital and geospatial technologies
I can work with layers and other features on geospatial technology platforms to further explore data and interconnections.

I can access and use geospatial technology platforms such as GIS
I can use geospatial technology platforms to explore data and find patterns.

I can construct paper maps using correct cartographic conventions
I can use a pencil, paper and ruler to construct a map that follows BOLTSS conventions.

I can interpret different map types using cartographic conventions
I understand data found in different types of maps and graphs and use the data to answer questions about place studies.

Digital and geospatial technologies

Spatial distributions and patterns

1 Describe the *distribution* of houses on the right-hand image of Source 1.

Patterns and interconnections

2 Source 1: List the factors that may be *interconnected* with the economic divide in Johannesburg.

3 Using a mind map, explore other factors that may cause differences in human wellbeing on a regional *scale*.

Changes and implications

4 Rank the following factors in order of importance for improving human wellbeing. Justify your ranking with examples.
- Access to safe and durable housing
- Access to luxury items
- Access to travel and transport

Communicate data

5 List two **primary methods** that could be used to collect data on the happiness and wellbeing of a community in a particular *place*.

Digital and geospatial technologies

6 Outline how geospatial technology could be used to assess human wellbeing within a *region*.

Why are Nordic people so happy?

Each year, the World Happiness Report ranks 156 countries according to their objective and subjective wellbeing. Since 2013, Finland, Denmark, Norway, Sweden and Iceland have all ranked in the top 10 happiest countries in the world. The high ranking of these Nordic countries is largely attributed to their government policies, strong sense of place, social and economic equality, and level of development.

Source 1

The Nordic region, also known as Scandinavia, is in northern Europe. It comprises Denmark, Norway, Finland, Iceland and Sweden.

The countries of northern Europe

Source: Rainer Lesniewski/Alamy

Given their proximity to the North Pole, Nordic countries tend to experience less seasonal variation when compared to other regions. Reykjavík, the capital of Iceland, is known as the 'land of the midnight sun' and throughout summer the Sun may only set for around 3 hours. Conversely, in winter the Sun may only rise for 4 hours and so it is largely dark throughout the 'day'. While researchers have found that, generally, people in warm, tropical environments do tend to be happier than those in colder regions, the happiness of people in warmer regions fluctuates as the weather changes.

Consider how you feel when we experience a string of warm sunny days during the school week followed by a wet, cold weekend. Nordic countries, while dark and cold during the winter, have generally consistent climate patterns and so weather has been found to have minimal impact on people's happiness in these places.

One of the clearest reasons for Nordic happiness appears to be government stability and reliable welfare benefits. Research has shown that populations are generally happier when they have access to a quality democratic government, generous welfare benefits, stable employment and clear rights. When compared to other nations, Nordic countries tend to rank in the top 10 for 'government quality', which is considered to contribute strongly to their overall happiness. The quality of their governments is also attributed to the local sense of freedom. Using long-term studies, researchers have found that a positive change in a population's sense of freedom can increase subjective wellbeing by up to 30 per cent.

Source 2

Finland has a wide range of services, infrastructure and open space for recreation.

A sense of equality also has been seen to increase overall life satisfaction in Nordic people. When rating our subjective happiness and life satisfaction, we tend to look to others and compare access to resources, wealth, education and quality of life. Consider how you compare yourself to others in your classroom when assessing your happiness based on who has a newer phone or higher test score.

Given the high levels of equality in Nordic countries, there is a lower margin of difference between the 'rich' and 'poor' to accessing job security, welfare and resources when compared to other MEDCs, such as the USA. As a result, people do not negatively compare themselves to others as much and thus whether people are wealthy or not has less influence over their overall happiness ratings.

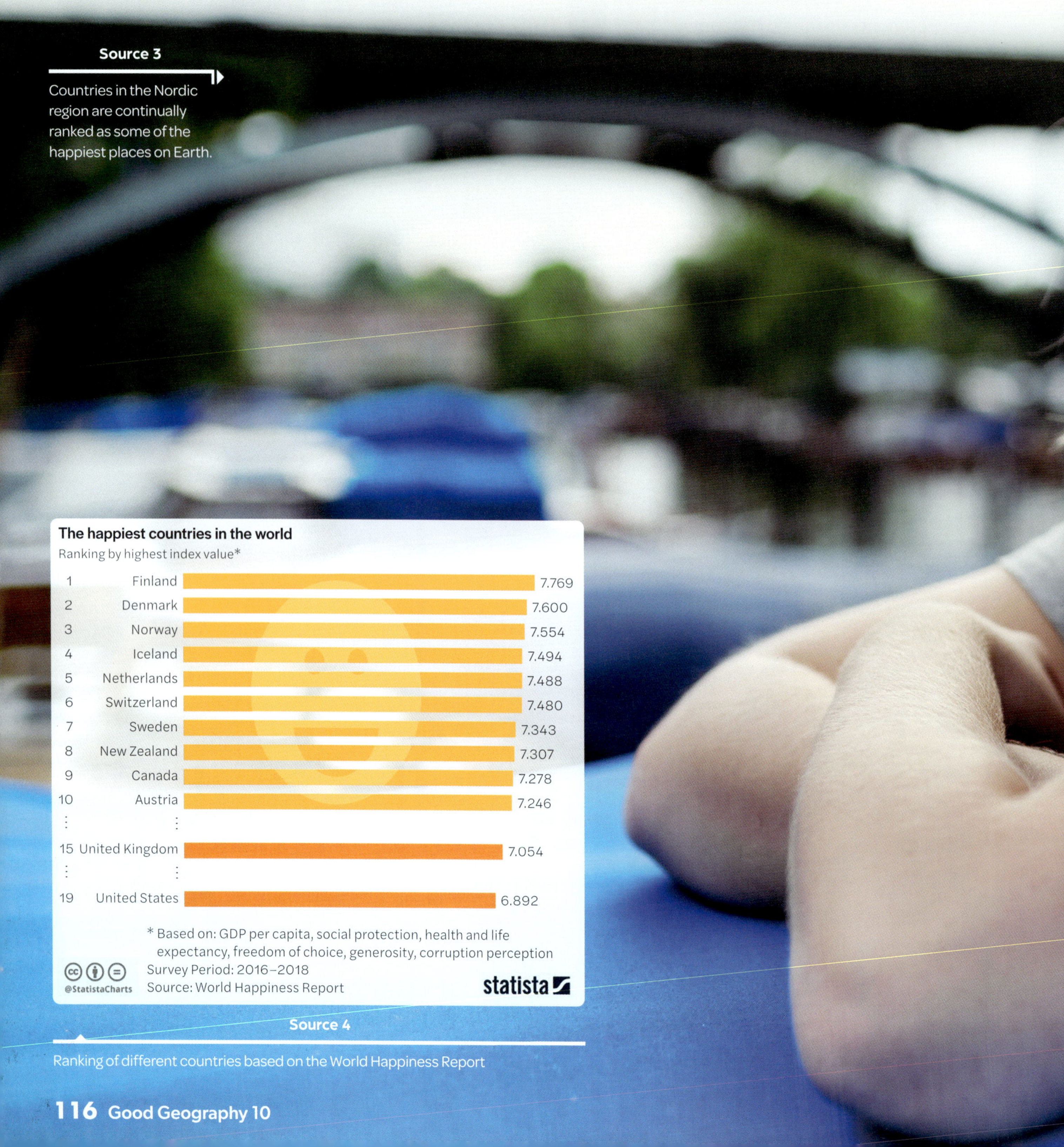

Source 3

Countries in the Nordic region are continually ranked as some of the happiest places on Earth.

Source 4

Ranking of different countries based on the World Happiness Report

Learning ladder G4.1

Show what you know

1 Outline the location of one Nordic country and state the absolute location of its capital city.

2 Create a list of five factors that influence people's happiness around the world. Create a photo essay that illustrates these factors in Nordic countries.

Spatial distributions and patterns

Step 1: I can identify spatial distributions and patterns

3 Source 4: Consider the top 10 happiest countries in the world. Using SHEEPT, identify key characteristics these *places* have in common to allow for such happy populations.

Step 2: I can use data to quantify spatial distributions and patterns

4 Create a comparison table between one Nordic country and one country from another world region. Research and create a statistics bank comparing happiness and wellbeing indicators for these locations. Identify and discuss any trends in the data with a partner.

Step 3: I can describe spatial distributions and patterns

5 Complete a PQE describing the relationship between GDP and average quality of life in different world *regions*.

Step 4: I can use data to support exceptions to spatial distributions and patterns

6 As a result of consistent seasonal patterns, climate was found to have little influence over happiness in the Nordic countries. Predict how global warming may influence happiness and wellbeing in this *region* over time. Where possible, use quantitative evidence or examples in your response.

HOW TO

PQE, page 154
SHEEPT, page 156
Graphing, page 161

How is Japan improving community wellbeing?

Japan is facing a huge demographic challenge: an ageing population coupled with potential population decline. Decades of low fertility rates, combined with one of the longest life expectancies in the world, means that, by 2025, a third of the Japanese population will be over the age of 65.

Year	Population	Growth rate	Fertility rate	Median age	Urban population
1970	105 million	1.28%	2.04	28.8	71.9%
1990	125 million	0.43%	1.65	37.3	77.3%
2010	128 million	0.03%	1.34	44.7	90.8%
2030	121 million	−0.5%	1.37	52.1	93.3%
2050	106 million	−0.7%	1.37	54.7	97.4%

Source 1

Japan's changing demographics

Source 2

The population of older Japanese citizens is increasing.

In 1945, Japan was a nation in ruins. The legacy of World War II's firebombing campaigns and the atomic bombings of Hiroshima and Nagasaki had left the country in tatters. Yet, decades later, it rose to become the world's second-largest economy and a vibrant hub of manufacturing and technological advancement that transformed the everyday lives of people and industries around the world. A post-war baby boom saw the population grow rapidly, but now it is in decline.

Factors that have contributed to the population challenge in Japan include:

- rural-to-urban migration has contributed to some of the highest rates of urbanisation and population density in the world
- cost-of-living pressures have made having a large family very expensive
- for many women, having children meant fewer work opportunities, essentially forcing them to choose between their career or having a family
- immigration rates have long been the lowest among MEDCs.

Impacts of Japan's population challenge include:

- as more people reach the age of retirement and fewer children are born, Japan's workforce is actually shrinking; simply put, there will not be enough workers to replace the current generation
- with fewer people in the workforce, fewer people will pay tax, meaning reduced revenue for governments, forcing them to slash their spending
- some industries, such as aged care and medical fields, continue to grow as the population ages, but many others, including schools in rural areas, will begin to decline. Businesses without enough workers will also close
- with a heavily urban population, many rural areas can fall into neglect, particularly as revenues and public spending decline
- with shrinking local industries, such as agriculture and manufacturing, the nation will become increasingly reliant on imports
- while government revenues are decreasing, it will face increased costs for healthcare and pensions.

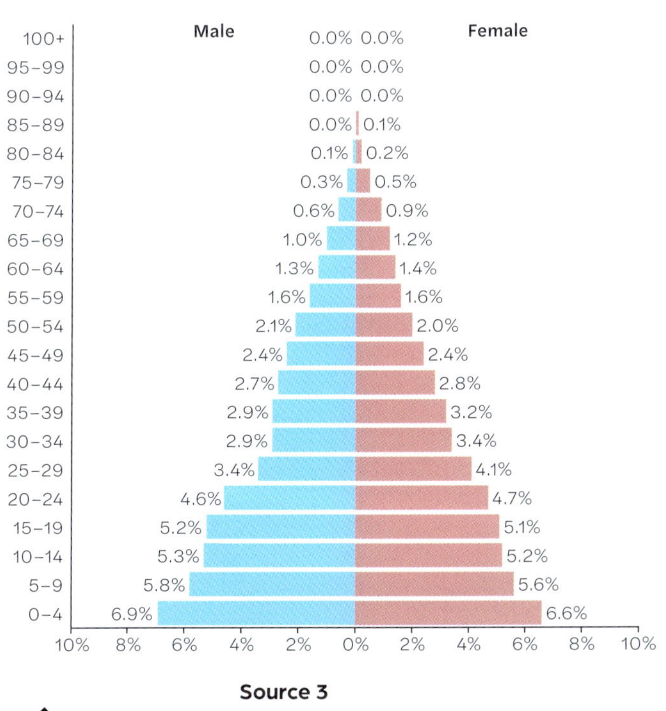

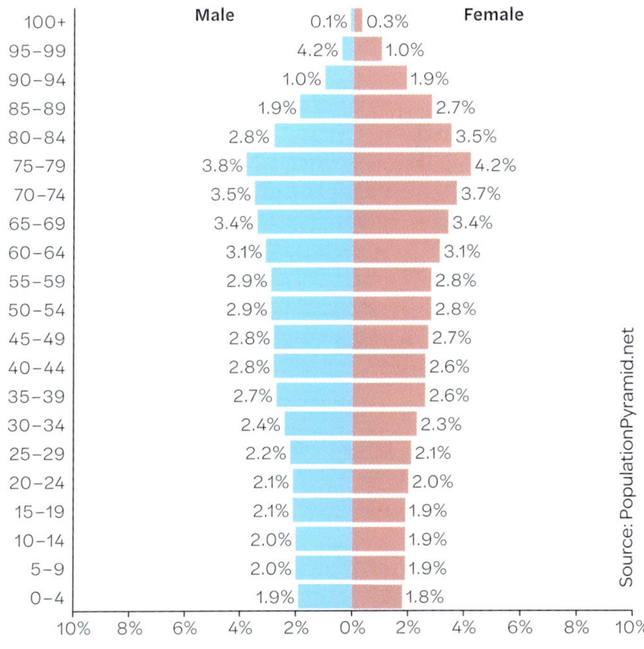

Source 3

Population pyramids of Japan in 1950 and projected for 2050

With these population challenges in mind, the Japanese government have created a range of responses to address these issues. At a policy level, the government has raised the eligibility age for the national pension from 60 in 2018 to 65 in 2025, with a view to further increasing it to 70 in the near future. This aims to increase both workforce capacity and taxation income, while also reducing the growth of pension payments.

Increasing female workforce participation is another way to maintain the economy, part of a broad economic stimulus package introduced by former Prime Minister Shinzo Abe, dubbed 'Abenomics'. Other responses aim to increase birth rates by supporting the costs of childbirth and providing funding for childcare and family support.

The growing cohort of elderly citizens remains a significant challenge to economic and social wellbeing and while a special category of work visa was created in 2016 to help fill work shortages – by 2025, the government predicts a shortfall of around 380 000 specialist workers – only 17 such visas were issued to candidates who passed difficult language exams. The development of elder-care robots has been flagged as a priority area for government funding, both to help fill the labour shortage and offer an economic lifeline for domestic industries to export the technology to other countries with similarly ageing populations.

Current technology available includes beds that transform into wheelchairs; robotic therapy animals; robots that assist in areas where power, mobility and monitoring are required; and robots that run games, coordinate exercise sessions and act as basic conversation partners. The Japanese labour ministry has invested more than $50 million in developing and implementing such technology across 5000 facilities nationwide.

Source 4

Robots are becoming part of elder care in many countries.

JAPAN

16% 

of the population live in relative income poverty

14% 

would be at risk of falling into poverty if they had to forgo 3 months of their income

17% 

of poor households spend more than 40% of their income on housing costs

There is no data available on life satisfaction

11% 

say they have no friends or family to turn to in times of need

There is no data available on satisfaction with time use

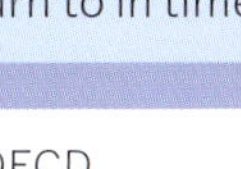

Source: OECD (2020), How's Life? 2020: Measuring Well-Being

Source 5

This OECD infographic shows deprivations in Japan for selected indicators of current wellbeing.

Learning ladder G4.2

Show what you know

1 What population challenge is Japan facing?

2 How will people's lives and wellbeing be affected by this challenge?

Changes and implications

Step 1: I can identify that changes occur in the characteristics of places over time

3 Using Source 2, describe the trends in Japan from 1970 to 2050 for:
- a population
- b growth rate
- c fertility rate
- d median age
- e urban population.

Step 2: I can describe how places have changed over time

4 a Source 3: Describe how Japan's working-age and non-working-age cohorts have changed over time. Ensure you use data to support your response.

b Suggest how population growth and economic growth have contributed to urbanisation.

Step 3: I can explain the causes behind the change over time in a place

5 To what extent is technology an adequate solution for Japan's population challenge? With a partner or in small groups, outline arguments for and against.

Step 4: I can make predictions and outline consequences of change over time

6 Immigration has helped other countries, such as Germany, maintain population growth despite an ageing population and has been touted as a possible solution for Japan's population challenge. Evaluate the following statement: 'Immigration would help maintain Japan's population growth, but would bring many other challenges with it'.

PQE, page 154
SHEEPT, page 156
Answering questions, page 162

How does conflict impact human wellbeing?

More than 40 per cent of people who live in extreme poverty are affected by conflict or corruption. War and violence impact people's wellbeing in a variety of ways. In the short term, both combatants and civilians are killed, and communities must withstand unsafe living environments and struggle to access resources, education and employment. In the long term, the impacts of war are not only detrimental to mental and physical health but cause geographical **displacement** and prolonged limited resource access, resulting in **malnourishment** and extreme poverty.

How has civil unrest impacted wellbeing in Syria?

Syria is located in the Middle East and is bordered by the Mediterranean Sea, Turkey, Iraq, Jordan, Israel and Lebanon. Syria is characterised by desert, mountains and the Jabal al-Druze volcanic fields. More than 80 per cent of Syria's water is sourced from the Euphrates river that starts in Turkey and flows into southern Iraq.

After people protested against ongoing poverty and their lack of freedom under the dictatorship of Bashar al-Assad, the government launched a brutal retaliation. This escalated into a civil war in 2011. As of 2018, over 465 000 Syrians had been killed in the conflict and the UN estimates that almost 6.7 million Syrian people are displaced.

Syria

Source 1

A political map showing Syria and its major cities

Source: YAY Media AS/Alamy

A recent study investigating the impacts of the ongoing conflict in Syria found that between 2008 and 2015 it became 31.7 per cent more difficult to access healthcare, and social support reduced by 36.0 per cent when compared to other eastern Mediterranean nations. This has resulted in one of the largest declines in life satisfaction and the researchers found that 'negative emotions' in Syria increased by 37.5 per cent. The long-term impact of the ongoing civil war in Syria is yet to be determined; however, it is clear that this civil conflict has global ramifications.

Source 2

The Euphrates River

Source 3

The Syrian civil conflict has devastated both human and natural environments.

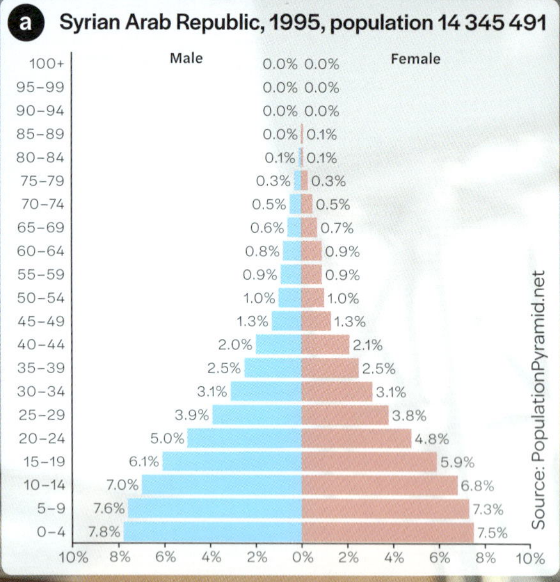

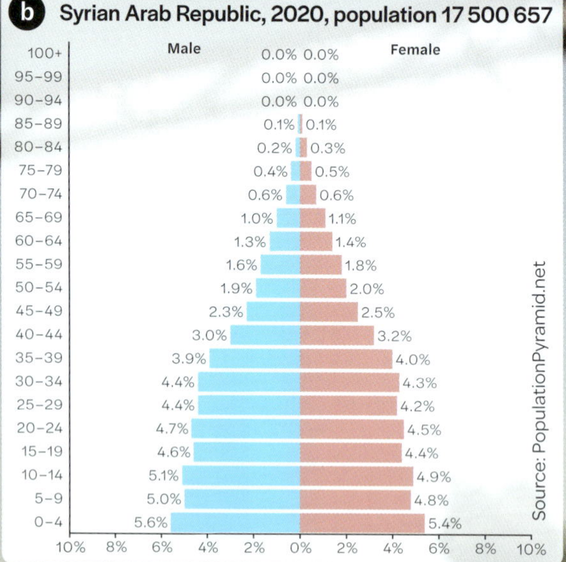

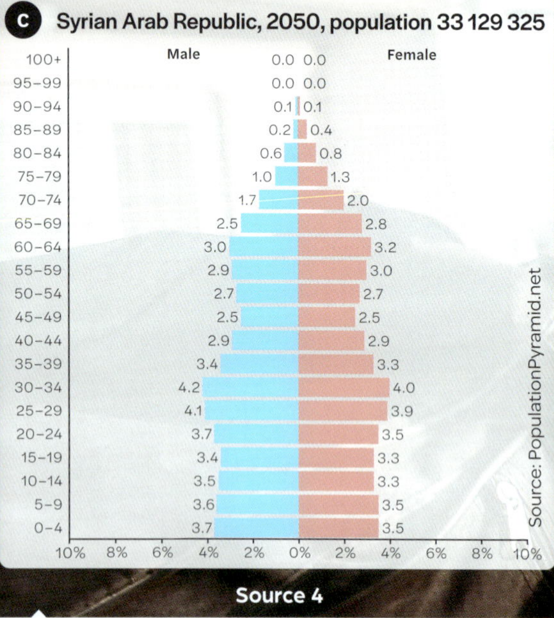

Source 4

Syria's population change from **a** 1995 to **b** 2020 and **c** predicted in 2050

The temporal change in the level of life satisfaction in 163 countries from 2006 to 2016

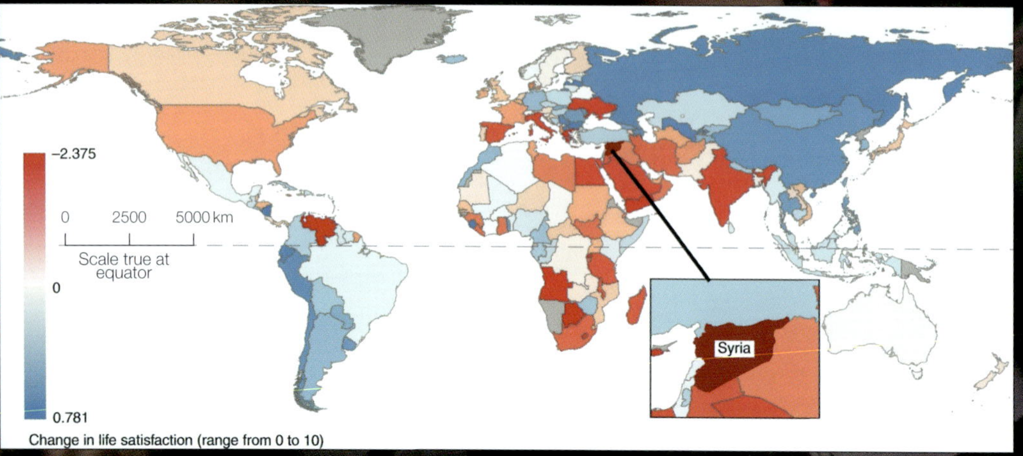

Source 5

Map showing the impact of the Syrian war on life satisfaction. Syria has had the most negative change of all countries.

Source: F. Cheung, A. Kube, L. Tay, et al. *Nature*

Source 6

As a result of the ongoing conflict, almost 40 per cent of the Syrian population are now refugees. These people are fleeing the region to seek safety and security in surrounding countries such as Turkey.

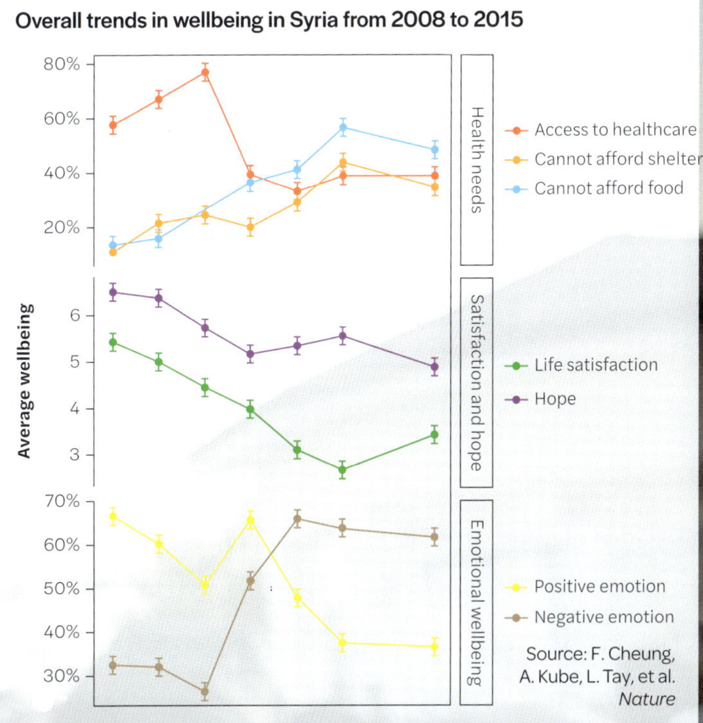

Overall trends in wellbeing in Syria from 2008 to 2015

Average wellbeing

Health needs: Access to healthcare; Cannot afford shelter; Cannot afford food

Satisfaction and hope: Life satisfaction; Hope

Emotional wellbeing: Positive emotion; Negative emotion

Source: F. Cheung, A. Kube, L. Tay, et al. *Nature*

Source 7

Syrian wellbeing trends for health needs, satisfaction and hope and emotional wellbeing from 2008 to 2014.

Learning ladder G4.3

Show what you know

1 Locate Syria on a world map and describe its location relative to Australia.

2 As a class, debate this statement: 'The long-term impacts of war are more detrimental to human wellbeing than the short-term impacts'.

Changes and implications

Step 1: I can identify that changes occur in the characteristics of places over time

3 List three *changes* that have influenced human wellbeing in Syria over time using data from the text (where possible) to support your answers.

4 Research images of Syria's natural biomes. Describe these *environments* and suggest how human activity has *changed* these locations over time.

Step 2: I can describe how places have changed over time

5 Source 4: Describe how Syria's population has changed over time.

6 Source 5: Compare Syria's life satisfaction rating with that of other countries in Asia.

Step 3: I can explain the causes behind the change over time in a place

7 Source 7: Explain the *interconnection* between the factors highlighted and the wellbeing of Syrians.

Step 4: I can make predictions and outline consequences of change over time

8 Source 7: Predict the future of the emotional wellbeing of Syrians based on the patterns highlighted in the graph.

HOW TO

PQE, page 154
SHEEPT, page 156
Answering questions, page 162

How does consumerism impact human wellbeing in the Philippines?

Consumerism can be defined as our desire to own products that exceed our basic human needs. People often want to keep 'up to date' and to own new technology, cars, clothes and gadgets. Consumerism can have both positive and negative effects.

Consumerism can be a positive phenomenon because spending money on goods and services means that industries and economies expand. Consumers are able to purchase the products they want, while industry employs people to make and sell those products, including in LEDCs.

However, consumerism can also have negative effects on social and environmental conditions. Many people employed in manufacturing in LEDCs work in unsafe conditions and do not earn enough to lift themselves out of poverty. In 2005, around 59 per cent of the world's resources were bought and owned by 10 per cent of the population, creating huge disparities between the haves and have-nots.

Our desire to have the latest consumer goods creates a huge amount of waste. The World Bank estimates that 1.3 billion tonnes of rubbish are generated annually and this is expected to almost double by 2025. Items such as cigarette butts, plastic food containers, drink bottles, straws and shopping bags are some of the largest contributors to global waste. It is estimated that plastic takes more than 450 years to break down and, while recycling rates are increasing, by 2050 research suggests that there will be more plastic in the ocean than fish!

Source 1

The beautiful beaches of the Philippines are often promoted to attract tourists.

Waste and human health in the Philippines

The dumpsite of Payatas, located in Quezon City, Philippines, is home to more than 200 000 people. Many adults and children in this region risk their lives to earn a very low wage, rummaging through the tip to search for recyclables and sellable items and living close to polluted waterways. The community is exposed to deadly diseases such as pneumonia and tuberculosis.

Globally, it is estimated that over 1000 children under five die per day from diseases linked to unsafe water and lack of sanitation. Further, the instability of waste sites in Payatas means that workers may be crushed by the rubbish.

To make matters worse, as China reduces its recycling capacity, there is less demand for products collected by the Payatas communities, meaning income for these families has reduced.

'Hard-up families live in Payatas as an option to survive,' Jess Far, child protection officer at UNICEF Philippines, said. 'These are the poorest of the poor. They come to Manila from the provinces looking for a better life but they find that life is more difficult here than they expected. A lot of children in Payatas end up out of school because of poverty.'

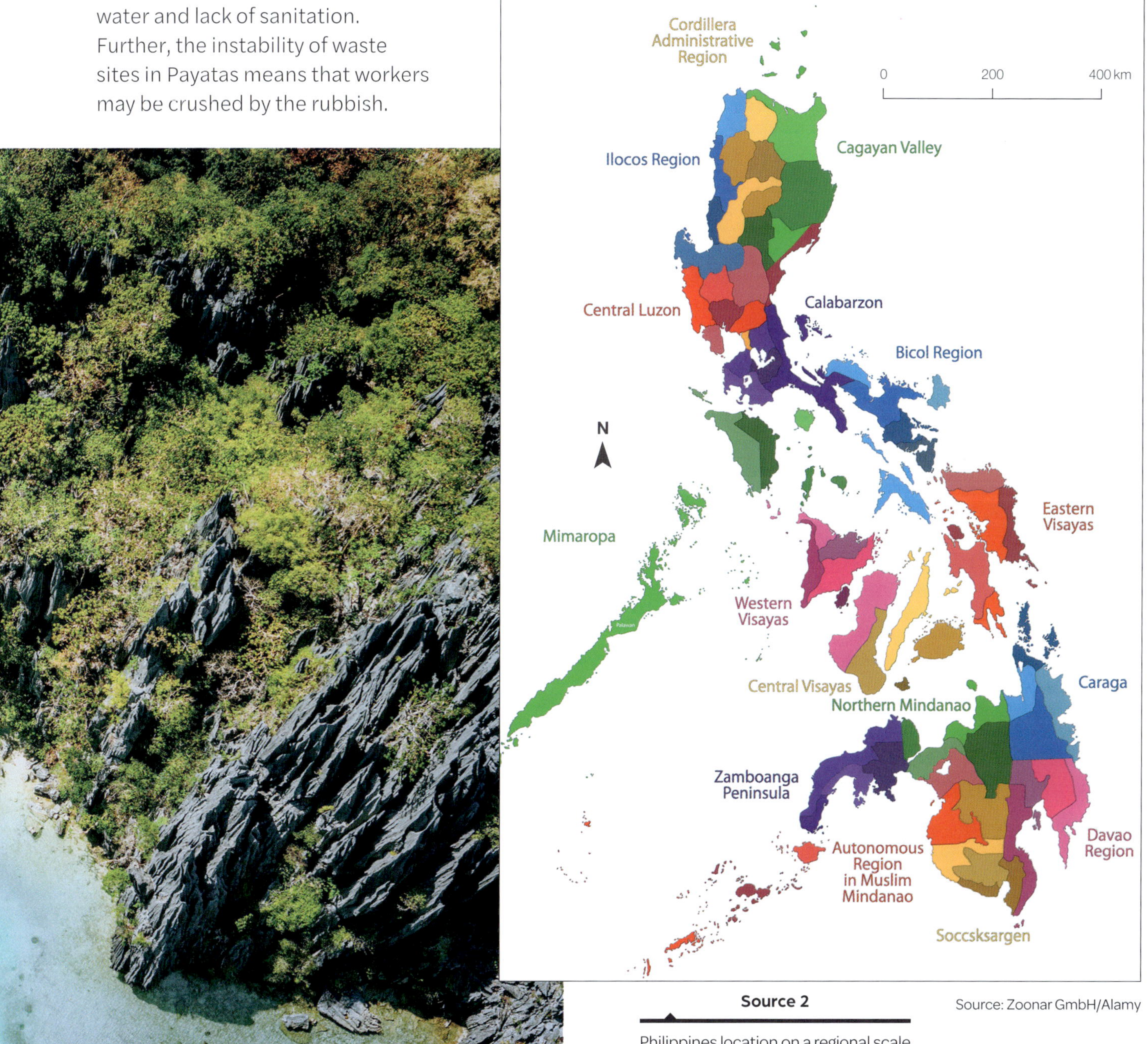

Source: Zoonar GmbH/Alamy

Source 2

Philippines location on a regional scale

'I collect and sell anything': Waste-picking a means to an end for Philippines' poor

By Shirley Escalante in Manila, ABC News website, 4 September 2016

Mr Lluz, 34, is a waste-picker, and has been for over 10 years. He lives with his wife Mercy and their five children in a community at the Payatas dumpsite. Along with some 6000 other waste-pickers in Payatas, Mr Lluz waits for the trucks that deliver their garbage haul from the city every day and then rummages through mounds of rubbish in search of recyclables — anything from used bottles, cans and newspapers to plastics, metals and tyres. Some Filipino waste-pickers choose to push carts through the streets of Manila to collect recyclables, while others still pick trash straight from bins. But Mr Lluz chooses to collect directly from the dumpsite, where he cleans the recyclables, organises them and then sells them. 'Bronze is the most expensive. I sell it for almost $US4 ($5.30) a kilo. Paper is the cheapest — it's worth two US cents ($0.03) a kilo,' he says. 'I collect and sell almost anything.'

For Mr Lluz, scavenging at the Payatas dumpsite is seen as a more financially lucrative job prospect in comparison to his days as a factory worker. 'I earn almost $US7 from waste-picking the whole day,' he said. 'My daily wage at the factory was only a little over $US4. And it's a fixed amount. I can earn more when I collect and sell recyclables. One can't be picky or lazy at the dumpsite.' Working at the dumpsite could be loathsome, and would never be considered an option for most — but for Mr Lluz, it has improved his way of life, and that of his family. 'Because I earn more, my family eats three meals a day. And I am able to send my five children to school,' he says.

Over 10 years later, Mr Lluz says he believes waste-picking will allow him to fulfil his simple dreams, such as putting a roof over his head and sending his five children to school. He also hopes to see his children complete a university degree one day — a dream he says he was not able to achieve because of economic difficulties. If a better job prospect came along, Mr Lluz says he 'just may grab it.' But for the time being, he says he is happy to stay working at the Payatas dumpsite.

Source 3

Children help parents collect recyclables from the dump to sell. Children are at increased risk of serious illness and injury as a result of these activities.

Item found in Payatas	Potential sell price *before* China reduced recycling capacity	Potential sell price *after* China reduced recycling capacity
Tin cans	25 pesos per kilo	15 pesos per kilo
Plastic cups	12 pesos	5 pesos
Clear plastic	2 pesos	1 peso

Source 4

Income earned by selling recyclables in Payatas before and after China closed many of its recycling plants

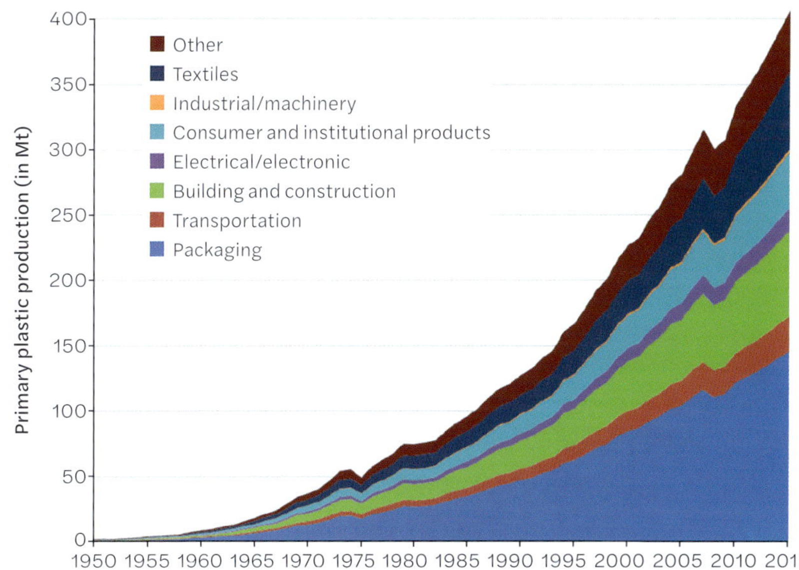

Source 5

Global plastic production per industry over time (1950–2015)

Source: Roland Geyer, Jenna R. Jambeck, Kara Lavender Law, *Science Advances*, 19 Jul 2017

Learning ladder G4.4

Show what you know

1 Describe the location and geographical characteristics of the Payatas dumpsite.

2 Source 4: Create a graph highlighting the *change* in income for locals at the Payatas dumpsite over time.

3 Discuss the following statement as a class: 'Local actions regarding waste have global consequences'.

Patterns and interconnections

Step 1: I can provide short explanations for patterns and interconnections

4 Source 5: Using SHEEPT, suggest reasons for the growth in overall use of plastics on a global *scale* since 1950.

5 Consider your response to Question 4. Outline the potential environmental and social consequences for this increase in plastic waste.

Step 2: I can explain patterns and interconnections

6 Referring to the sources, predict how human wellbeing and the natural *environment* may be affected by waste disposal in the future.

Step 3: I can use data to support explanations of patterns and interconnections

7 Using examples or data from the text, discuss how consumerism is linked to human wellbeing and happiness in different world *regions*.

8 Describe the interconnections between environmental health and human wellbeing using evidence from this spread to support your response.

Step 4: I can use relevant sources to research further reasons for patterns and interconnections

9 While working at the Payatas dumpsite is dangerous for human health and safety, it seems like subjective happiness is linked to hope for the future in this *place*. Discuss how human perspective may influence wellbeing and happiness despite economic or environmental turmoil.

SHEEPT, page 156
Answering questions, page 162

How does mining impact human wellbeing in Mongolia?

In many locations around the world, use of natural resources is closely tied to economic growth and wellbeing. The impacts of this can be both positive and negative. The growth of industry and creation of jobs and opportunities can often be tempered by environmental degradation and the issues associated with rapid urbanisation.

Mongolia produces more than 10 per cent of all global coal, as well as significant reserves of copper, gold and other minerals. With ready buyers available in neighbouring China, this wealth of natural resources has fuelled significant economic growth and helped improve the standard of living in the nation. Yet development has come at a cost, as rapid industrialisation has greatly increased pollution levels, a problem further exacerbated by rural–urban migration and urbanisation in the traditionally nomadic nation.

The capital city, Ulaanbaatar, has doubled in population over the past two decades, from around 765 000 in 2000 to more than 1.5 million residents in 2020. Mongolia is one of the world's least densely populated nations, with an average population density of just two people per square kilometre, but in Ulaanbaatar this figure now exceeds 320 people per square kilometre and is rising rapidly. High-density apartments now dot the landscape as a the city tries to house new migrants.

The World Health Organization (WHO) has identified Ulaanbaatar as one of the most polluted cities on the planet, with dangerous levels of fine particulate matter (PM2.5) in the air. PM2.5 levels in Ulaanbaatar can reach 133 times the WHO recommended level, particularly during the bitter winters, as citizens in traditional housing areas and the urban fringe burn coal to keep warm. In May 2019, the Mongolian government banned people burning unprocessed coal to try to mitigate the issue.

Positives	Negatives
• Significant increases in key development indicators, including life expectancy, years of schooling and gross national income per capita • Increased spending and investment in public transport, schools and medical facilities • Improvements in maternal health and educational opportunities for women	• Loss of arable land due to urban growth • Use of water in mining in semi-arid areas • Generation of dust from mining trucks • Increased siltation of rivers and water sources • Traffic congestion • Air pollution from industrialisation and burning of fossil fuels

Source 1

Positives and negatives of Mongolia's development

Trends in Mongolia's HDI component indices 1990–2019

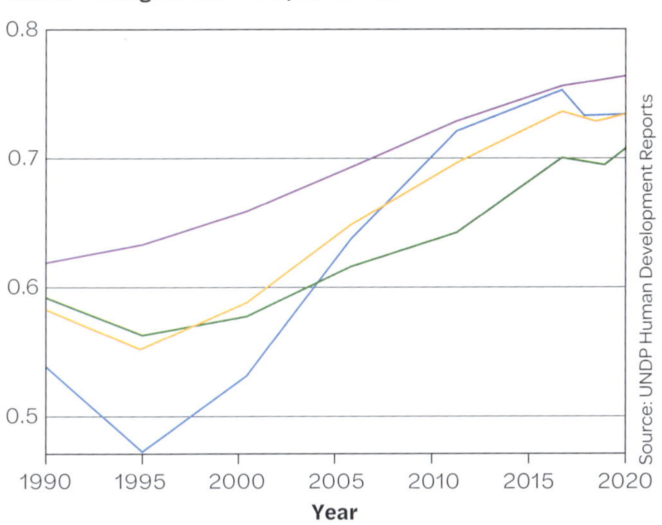

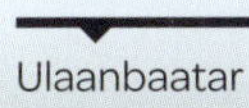

Source 2

Many of Mongolia's Human Development Index measurements are improving.

Source 3

Ulaanbaatar

Learning ladder G4.5

Show what you know

1 Identify the primary industry that has driven Mongolia's economic growth over the past two decades. *Interconnection* with which country has facilitated this?

2 'The positive aspects of Mongolia's development outweigh the negatives.' Debate as a class or prepare a statement for or against.

Changes and implications

Step 1: I can identify that changes occur in the characteristics of places over time

3 Rank the positive and negative impacts of *change* in Ulaanbaatar. Discuss your rankings with a partner.

Step 2: I can describe how places have changed over time

4 Using Source 2, describe the trend in human development between 1990 and 2018.

Step 3: I can explain the causes behind the change over time in a place

5 Prepare a cause-and-effect diagram for Ulaanbaatar. Using Source 1, suggest specific factors to demonstrate *interconnection*.

Step 4: I can make predictions and outline consequences of change over time

6 'Future population growth in Ulaanbaatar is going to have more negative than positive impacts for the people living there.' To what extent do you agree with this statement? Discuss using evidence and examples.

HOW TO

PQE, page 154
SHEEPT, page 156
Answering questions, page 162

How does human wellbeing vary between communities in southern Africa?

Southern Africa has been described as a land of contrasts. From Namibia's hyper arid Skeleton Coast in the southwest to the stunning white sands and tropical resorts of Mozambique to the southeast, the region encompasses a myriad of biomes, peoples, resources and infrastructure. In terms of human wellbeing, there is also considerable variation. Many nations in the region rank low on the Human Development Index, but at a local scale there are significant differences in terms of both wellbeing and development.

In geographic terms, a range of factors affect the wellbeing of the people of southern Africa. Economic factors, including income, industry and employment, play a major role. In the island nations of Seychelles and Mauritius – where GDP per capita is among the highest on the continent – tourism offers significant opportunities for work, training and education. In Mozambique, which also has significant potential for tourism, a long-lasting civil war (from 1977 to 1992) caused significant damage to infrastructure, as well as an exodus of wealthy citizens to Portugal; civil war has also been a problem for Angola.

Source 1

Up-market housing in a gated community in Johannesburg, South Africa

Source 2

A township in Soweto, Johannesburg, South Africa

The events of the past also play a role in the disparity in modern South Africa, where the shadows of the **apartheid** era still linger. This is evident in the city of Johannesburg, where townships in Soweto, on the city's fringe, stand in stark contrast to gated communities, upmarket suburbs and new developments such as Steyn City.

In neighbouring Botswana, an abundance of high-quality diamonds drives economic growth, supporting the development of infrastructure and political stability. Zimbabwe, located to the northeast of Botswana, has a wealth of natural resources and was once known as 'the breadbasket of Africa', but an autocratic government and **hyperinflation** have brought the country's economy to its knees. Across many nations, the legacies of colonialism are also evident, with the Portuguese, English, Germans and Dutch all having influenced the local populations.

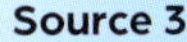

Source 3

The sandy beaches of Mauritius

The nations of southern Africa

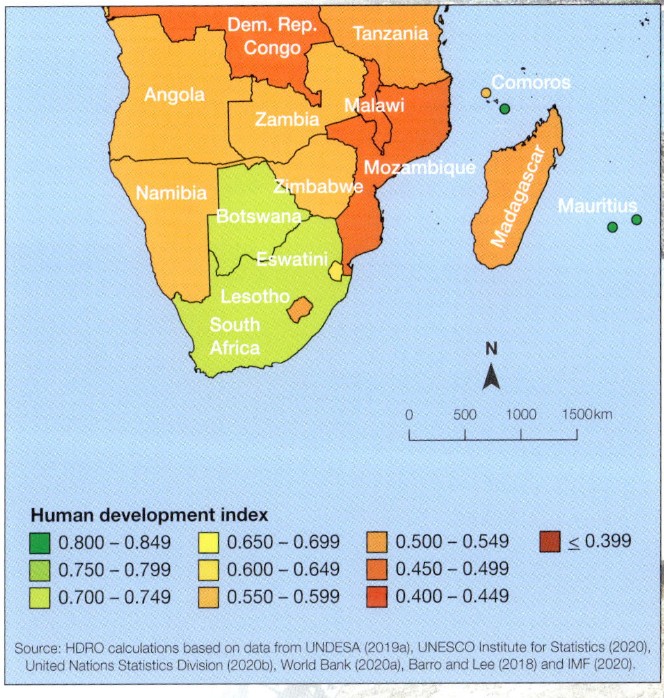

Source 4

The countries of southern Africa and their progress measured against the HDI

Source: Wikimedia adapted from Asus2004

Source 5

Maputo, the capital of Mozambique

Beyond the broad trends, myriad factors affect the differences in wellbeing between communities across southern Africa, including: opportunities for women, foreign investment, natural resources, level of corruption, freedom of the press, net migration rates, urbanisation, land degradation, provision of public health services, rates of HIV infection, rates of malaria, internal migration and more. The high degree of interconnection between factors creates a range of challenges to support wellbeing among communities.

Source 6

The savannah in Malawi

Source 7

GDP of some southern African nations, in International dollars

Country	GDP per capita (PPP)	% of GDP from agriculture	% of GDP from industry	% of GDP from services
Seychelles	$30 517	2.3	11.0	72.3
Mauritius	$23 882	2.9	17.3	67.7
Botswana	$18 553	1.9	28.3	58.2
South Africa	$13 034	1.9	26.0	61.2
Namibia	$10 064	6.6	26.6	59.3
Eswatini	$9003	8.5	34.1	52.1
Angola	$6966	6.7	54.4	39.3
Lesotho	$2824	5.3	29.1	54.6
Zimbabwe	$2961	8.3	20.6	61.3
Madagascar	$1720	23.3	17.2	52.1
Mozambique	$1338	26.0	22.9	39.9
Malawi	$1107	25.5	12.9	54.4

Source: World Bank, 2019

Source 8

Zimbabweans protesting against former President Robert Mugabe in 2017

Learning ladder G4.6

Show what you know

1 Consider the different images in this section. What clues do they provide about the diversity of southern Africa?

2 Source 7 divides GDP into three broad categories: agriculture, industry and services. What kind of areas does each cover? Discuss with a partner.

Patterns and interconnections

Step 1: I can provide short explanations for patterns and interconnections

3 Create a SHEEPT chart to classify the factors that influence disparity across southern Africa. How many different examples can you come up with?

Step 2: I can explain patterns and interconnections

4 Using the SHEEPT chart you've created, rank the factors in terms of their influence on a southern African nation of your choice. Which factors have the greatest impact on wellbeing? Why?

Step 3: I can use data to support explanations of patterns and interconnections

5 Source 7: Respond to the following statement: 'In nations where GDP per capita is higher, the economy is less reliant on agriculture'.

Step 4: I can use relevant sources to research further reasons for patterns and interconnections

6 Conduct further research into Steyn City at http://mea.digital/GHV10_G4_1 and Soweto. What differences do you notice between both? What challenges and opportunities do the residents face? Suggest how this might impact on their wellbeing.

7 Select a nation other than South Africa. What examples can you find of disparity there? Prepare a visual presentation to share with your peers.

SHEEPT, page 156
Answering questions, page 162

G4.7

getting involved

How do I make a difference?

On one level, clothing can be considered as largely practical. Coats keep us warm in the winter, and sandals stop our feet burning on hot sand. But clothing can also make a statement about who you are, what you stand for or what mood you're in. Many careers are centred around designing and wearing clothes, dressing people and influencing popular fashion choices. One thing people often do not consider is the impact of clothing on the environment.

Clothing can be a huge source of waste. Australians have been calculated to buy around double the global average of textiles each year at around 27 kilograms; however, we then throw away up to 23 kilograms, which ends up going to landfill. While having a clear out and donating to a charity seems more environmentally friendly, around 25 per cent of op-shop donations also end up in the bin because clothes are damaged, marked or not suitable for use.

Over the past two chapters, we considered how health, wealth and education can influence human wellbeing; however, happiness is also linked to warmth, comfort and pride in yourself. Clean, up-to-date clothing can allow us to fit in with peers, access job opportunities and regain confidence and independence after a crisis.

Source 1

In Australia, 3.24 million people or one in eight adults and one in six children live below the poverty line. Charities like Fitted for Work assist those in need to access basic resources such as appropriate clothing.

fitted for wOrk

Championing empowerment through job readiness support.

When you're Fitted for Work, you're fitted for life.

Source 2

Fitted for Work aims to 'help women in Australia experiencing disadvantages get work, keep work and navigate through working life with success'. Through donations, it provides professional clothing, job training and a range of workshops to 'anyone who identifies as a woman and those who do not wish to be limited by definition'. In this way, Fitted for Work aims to break the poverty cycle by giving people the opportunity to succeed through both what they wear and the provision of mentoring and support.

More than 3 million Australians live below the poverty line, including 774,000 children.

Most of those affected are living, on average, $282 per week below the poverty line

Source 3

Poverty in Australia

	SINGLE	FAMILY
THE POVERTY LINE:	$457 INCOME PER WEEK	$460 INCOME PER WEEK
	13.6% or 1 in 8 people	17.7% or 1 in 6 children

THE POVERTY GAP: The average gap between incomes and the line is: $282 INCOME PER WEEK

Source 4

Thread Together, which was founded in 2012, aims to 'deliver new, good quality clothing and shoes to people in our community who are doing it tough'. Thread Together works with more than 500 fashion brands to save brand-new clothing from going to landfill and providing these items to thousands of Australians in need via a network of over 500 established charities and social service agencies. It has diverted more than 2 million units of clothing from landfill and provided new wardrobes to nearly half a million people.

Getting involved

Creating a charity drive for your school

1 Explore different charities online that assist those suffering from financial crises or homelessness in your local *region*. Choose one to focus on in more detail.

2 What are the main aims of this organisation? Who is it aiming to assist? Does it have any advertised targets such as fundraising goals or donations to meet?

3 Develop a list that highlights how the charity would like the community to be involved in its services. Does it need volunteers, clothing or financial donations?

4 Review the charity's past work and success stories. Evaluate the effectiveness of the work this charity undertakes in your local community.

5 Write a proposal to your principal suggesting how your school could get involved with this particular charity. How could you raise money, collect clothing or food donations or volunteer in a responsible and safe manner?

6 Create a series of posters advertising approved events, fundraisers or drives as per your proposal.

7 Carry out your suggested fundraising activities and write a short paragraph reflecting on how your participation helped improve local human wellbeing and happiness.

Masterclass

Learning ladder

Work at the level that is right for you or level-up for a learning challenge!

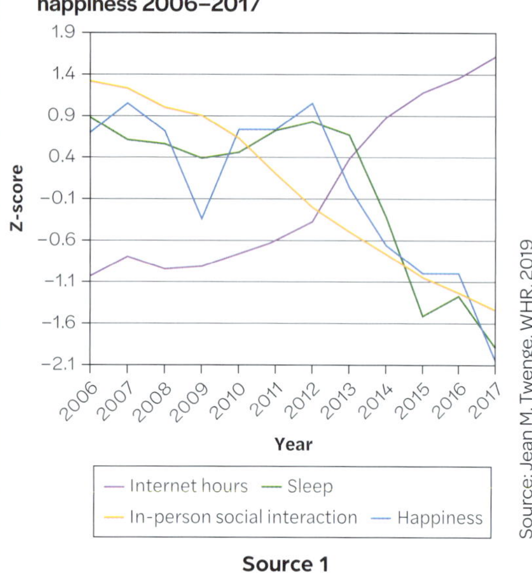

Source: Jean M. Twenge, WHR, 2019

Source 1

Standardised (z) scores for Years 8 and 10 students in the US, measuring happiness and time spent on the internet, sleeping and interacting with

Poverty rates in the USA, 2017

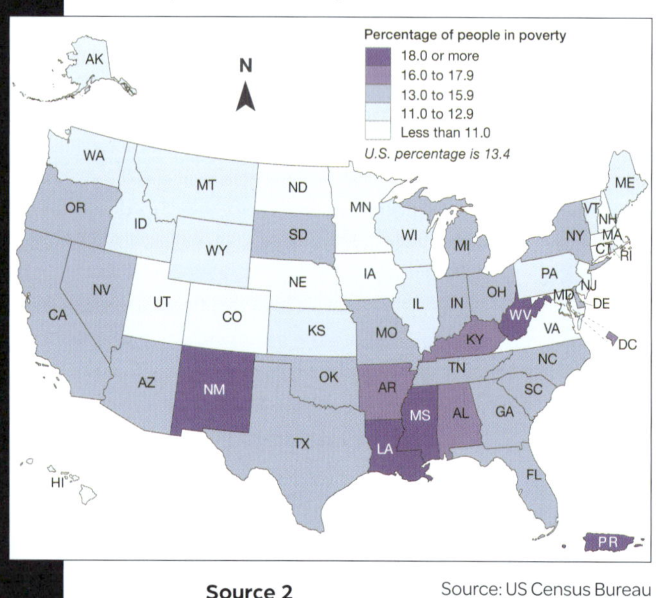

Source 2

Source: US Census Bureau

Poverty rates in the USA in 2017

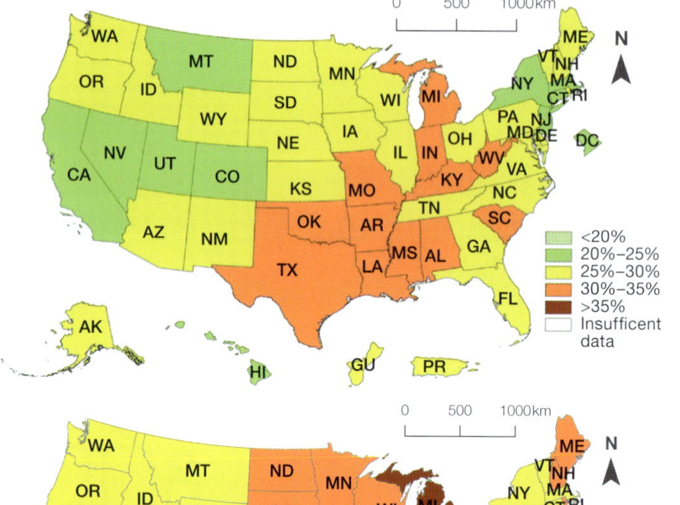

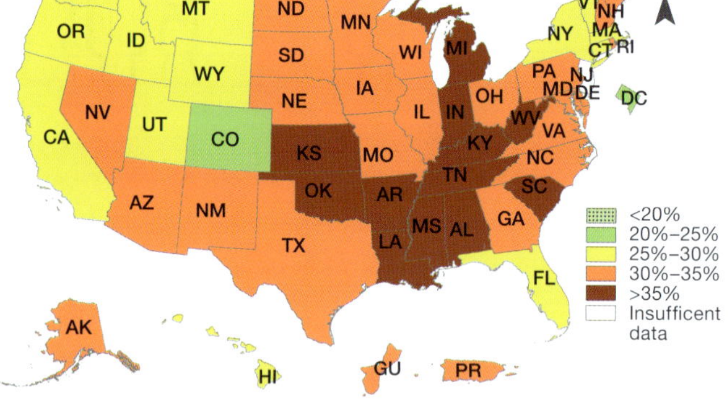

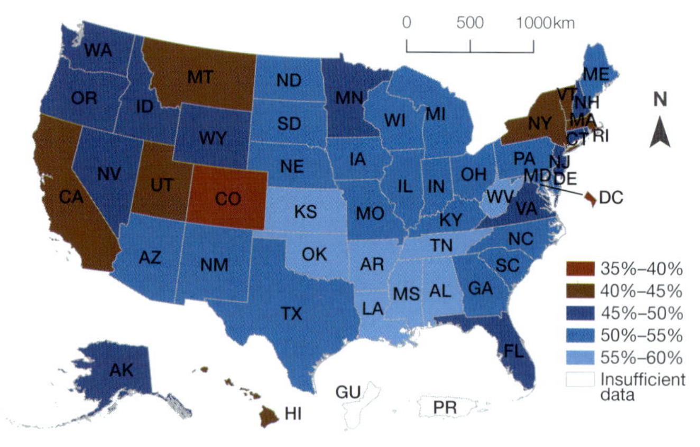

Source 3

Change over time in obesity rates by state in the US

Source: US Centers for Disease Control/ Custom Mapping Services

Step 1

a I can identify spatial distributions and patterns

Source 2: Identify one US state that has a high poverty rate.

b I can provide short explanations for patterns and interconnections

Source 1: Explain why internet usage may have increased in Years 8 and 10 students over time.

c I can identify that changes occur in the characteristics of places over time

Source 3: Identify how obesity rates in the US are changing over time.

d I can list primary and secondary methods useful for my study

Suggest how population data collections such as a census could assist in gaining data on human wellbeing at a certain *place*.

e I can interpret different map types using cartographic conventions

Source 3: Describe what these maps are showing and how the colour legend helps show *change* over time.

Step 2

a I can use data to quantify spatial distributions and patterns

Source 2: State the poverty rate of Texas.

b I can explain patterns and interconnections

Using SHEEPT, explain the main trend observed over time in Source 3.

c I can describe how places have changed over time

Source 1: Using PQE, describe how happiness ratings have *changed* in Years 8 and 10 students over time in the US.

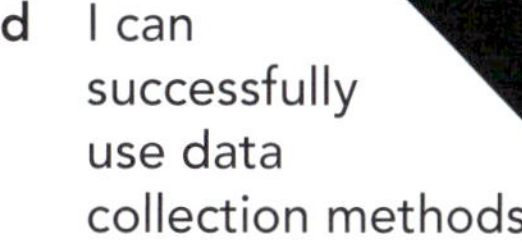

d I can successfully use data collection methods

Research poverty rates in another country of interest. How do they compare with the US?

e I can construct paper maps using correct cartographic conventions

State which elements of BOLTSS are missing from Source 3.

Step 3

a I can describe spatial distributions and patterns

Source 3: Describe the *change* in obesity rates in Colorado.

b I can use data to support explanations of patterns and interconnections

Source 3: Explain how obesity rates may be used to investigate wellbeing in the US.

c I can explain the causes behind the change over time in a place

Source 2: Using SHEEPT, explain why some states may have higher poverty rates than others. What can we infer about the general level of human wellbeing in those states as a result?

d I can filter collected data

'There appears to be a strong relationship between decreased happiness and increased internet usage in young people.' Justify this statement with data from Source 1.

e I can access and use geospatial technology platforms such as GIS

Access http://mea.digital/GHV10_G4_2. Is this data an objective or subjective measurement of human wellbeing in the US?

Masterclass

Step 4

a I can use data to support exceptions to spatial distributions and patterns

Source 1: 'Patterns of happiness were quite consistent from 2006–2012 with the exception of 2009.' Use data to support this statement and research an event that may have contributed to this sudden *change*.

b I can use relevant sources to research further reasons for patterns and interconnections

Source 1 suggests that a lack of sleep and human interaction can lead to a decline in happiness. Locate some reliable studies or sources that support this correlation.

c I can make predictions and outline consequences of change over time

Source 3: Identify and quantify three states that are predicted to see increased rates of obesity in 2030.

d I can organise data collected according to relevance for a research question

Rank sources 1–3 in order of relevance for understanding human wellbeing in the US. Justify your ranking.

e I can manipulate data using digital and geospatial technologies

Access http://mea.digital/GHV10_G4_2. Explore the interactive map and locate the following:

- i the US county with the highest death rate
- ii the US county with the lowest alcohol consumption rate
- iii the US county with the highest daily male smoking rate.

Step 5

a I can identify multiple spatial distributions and patterns

Describe the *spatial association* between obesity and poverty rates in the US.

b I can interpret causes of patterns and interconnections

Source 1: Discuss the relationship between internet usage, in-person contact and happiness in young Americans.

c I can interpret data to quantify predictions based on research

Source 1: Based on your experience during the COVID-19 pandemic, predict how internet usage, happiness and sleep patterns may have *changed* in young Americans.

d I can evaluate the success of research methods

Evaluate how successful three sources are at providing an insight into human wellbeing in the US.

e I can draw conclusions from geographical information in digital and geospatial technologies

Access http://mea.digital/GHV10_G4_2, then consider the statement, 'To understand human wellbeing, researchers should focus on health data'. To what extent do you agree with this statement? In your response refer to both objective and subjective measures of wellbeing.

Capstone

How can I understand place studies of human wellbeing?

In this chapter, you have learned a lot about human wellbeing. Now you can put your new knowledge and understanding together for the capstone project to show what you know and what you think.

In the world of building, a capstone is an element that tops off a building or a wall. That is what the capstone project will offer you, too: a chance to top off and bring together your learning in interesting, critical and creative ways. You can complete this project yourself, or your teacher can make it a class task or a homework task.

mea.digital/GHV10_G4

Scan this QR code to find the capstone project online.

Fieldwork G5

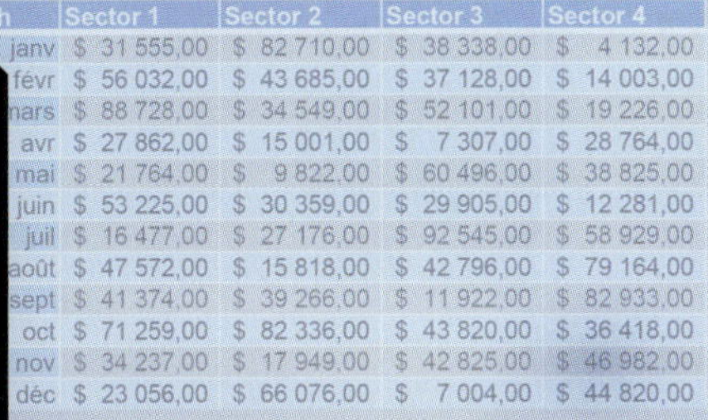

onth	Sector 1	Sector 2	Sector 3	Sector 4
janv	$ 31 555,00	$ 82 710,00	$ 38 338,00	$ 4 132,00
févr	$ 56 032,00	$ 43 685,00	$ 37 128,00	$ 14 003,00
mars	$ 88 728,00	$ 34 549,00	$ 52 101,00	$ 19 226,00
avr	$ 27 862,00	$ 15 001,00	$ 7 307,00	$ 28 764,00
mai	$ 21 764,00	$ 9 822,00	$ 60 496,00	$ 38 825,00
juin	$ 53 225,00	$ 30 359,00	$ 29 905,00	$ 12 281,00
juil	$ 16 477,00	$ 27 176,00	$ 92 545,00	$ 58 929,00
août	$ 47 572,00	$ 15 818,00	$ 42 796,00	$ 79 164,00
sept	$ 41 374,00	$ 39 266,00	$ 11 922,00	$ 82 933,00
oct	$ 71 259,00	$ 82 336,00	$ 43 820,00	$ 36 418,00
nov	$ 34 237,00	$ 17 949,00	$ 42 825,00	$ 46 982,00
déc	$ 23 056,00	$ 66 076,00	$ 7 004,00	$ 44 820,00

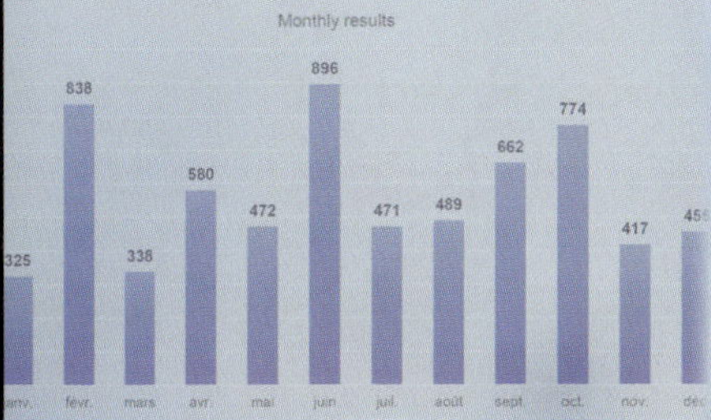

HOW DO I CONDUCT AND REPORT MY FIELDWORK?

page 142

How do I conduct and report my fieldwork?

You may have already conducted fieldwork as part of your geographic studies. Perhaps this fieldwork involved conducting surveys, completing sketches or answering structured questions about a particular topic. As you move towards VCE, fieldwork reports will become more formal and may involve more independent thought and research. Furthermore, your fieldwork will not always need to be conducted outside, but instead can be completed virtually using a range of reliable secondary sources.

Creating a research question

A **research question** is an overarching idea you want to investigate. Ideally, a research question will investigate a current gap in our knowledge or, in other words, look into a process or phenomena that has not yet been explored in academic or geographic literature. Once you develop a research question, you will need to write a hypothesis. A hypothesis is an 'educated guess' about what the answer to the research question might be. In this chapter, we will guide you to create your own research question and hypothesis and then to design fieldwork to answer your research question.

Choosing a fieldwork location

Choosing the right fieldwork location is vital. The place you choose will depend on your topic, research question and hypothesis. Make sure you confirm the location of your fieldwork before you begin to conduct research and collect data. You may decide that the topic you have chosen to investigate can be completed virtually. This will mean you will need reliable internet and devices to collect data from online sources.

Pre-fieldwork research

Before you start your fieldwork, do some background research so that you understand your chosen field site, its characteristics and any other relevant information. The internet is the most obvious and accessible source of information; however, seek out reliable websites. Remember: blogs, social media and other public-based sites may be biased and misleading.

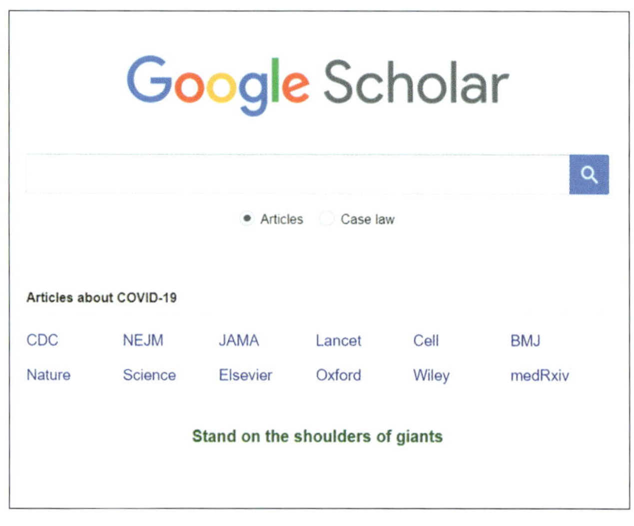

Source 1

Google scholar is a good starting point

In the past, you may have relied on government or education sites and books for reliable data. In addition to these, you may also wish to start exploring academic, peer-reviewed articles to gain more up-to-date and advanced data sets. When professional geographers and other scientists complete research, they write articles (similar to fieldwork reports) and submit them to academic journals. Peers or other professionals who are in the same field then review their work and provide suggestions on how they could improve their research process or add to their analysis. Once approved, the peer-reviewed article is published for other geographers and scientists, such as yourself, to read and use in their own investigations. While sites such as Google Scholar are good starting points, you may want to talk to your librarian about the kind of access your school has to peer-reviewed journal articles.

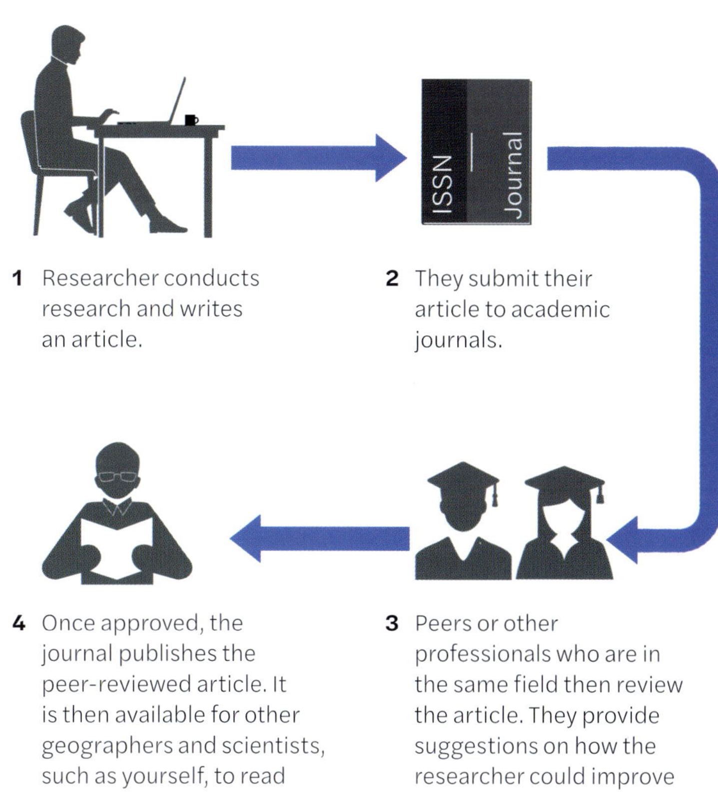

Source 2

The peer review process

Citing your research

While bibliographies are important to show your reader the source of your information and reduce the risk of plagiarism, as you move towards being a senior geographer you may need to start completing in-text citations and reference lists as part of your work. While different schools require different styles of citation, the basic idea is still the same.

Follow the general guide below for how to complete in-text citations.

1 When you have found some research you would like to include in your fieldwork report, note the information shown in Source 3.
2 Use the data and adapt any information you wish to include into your own words.

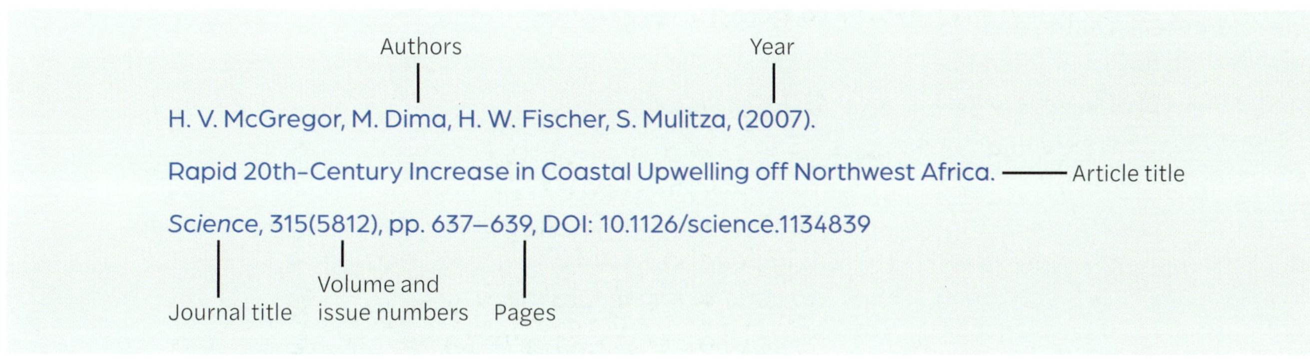

Source 3

Note down reference information early. Note: DOI (Digital Object Identifier) is a unique number to identify a particular publication.

3 At the end of the sentence or paragraph where you have used a study's or website's data or information, before you conclude with a full stop, insert a set of brackets; for example, 'The current Australian population exceeds 25 million people (...).'

4 Inside the brackets you need to list the last names of the author or authors who provided the information or the name of the company or government body, if the author names are not available, followed by the year this information was published. For example, 'The current Australian population exceeds 25 million people (Australian Bureau of Statistics, 2020).'

5 You then need to complete a full reference list containing all your cited references at the end of your report. 'Australian Bureau of Statistics (2020), Population Clock, accessed: 12 November 2020, https://www.abs.gov.au/ausstats/abs%40.nsf/94713ad445ff1425ca25682000192af2/1647509ef7e25faaca2568a900154b63?OpenDocument.'

The in-text citations act as little bookmarks to indicate to your reader where your information is sourced from, or where to find out more. The reference list at the end provides the full list of references for the reader to be able to find the location of the studies or data you used and complete their own investigations.

Writing an introduction

Where do you start when writing a fieldwork report? Your introduction provides a background to what you are investigating and why it is important. In other words, it provides context for your research. Remember:

- A formal fieldwork report does not use pronouns such as 'me' or 'I'.
- Cite research to show where you sourced information and to demonstrate you have used reliable sources.

You may think of your introduction like a funnel, where you start with the broadest ideas and finish with your research question and hypothesis.

Your opening lines may need to define some key terms or provide background information about your research topic. For example, this opening paragraph may begin with:

'Climate change is a natural process that has been occurring over hundreds of thousands of years (Smith, 2010) ...'

Your second paragraph starts answering your research question more specifically. You may need to provide some background secondary data or evidence of previous research. For example:

'In 2017 alone, Australians emitted over 16.96 tonnes of carbon per capita (Ritchie and Roser, 2020) and this is having major impacts on environmental change on both a national and global scale.'

Your third paragraph should be much more targeted towards your specific research area. For example:

'Coral reefs are one example of a natural environment that is particularly susceptible to global warming. Management of these sites is vital to the continuation of marine ecosystems and fragile food webs.'

Your final paragraph may include your research question and hypothesis. You may wish to explore why your research is important or how it fills a 'gap' in academic knowledge or understanding, such as:

'Therefore, it was questioned, "How can coral reefs be better managed to reduce the impacts of global warming?"'

Method	Primary or secondary?	Description	This data will help me answer my research question because ...	How will I gather the data?
Photos	Primary	I will take five photos of infrastructure, natural environments and recent developments in the field site using a digital camera.	Taking photos provides a visual representation of the human and natural environments present at the research site.	Using a digital camera, I can take photos that can then be uploaded to my report.

Source 4

Record your research methods clearly and consistently so that you can repeat them.

Reporting methodology

Conducting fieldwork involves collecting data. Primary data collection methods are tasks that you will do in the field to find data to answer your research question, and prove or disprove your hypothesis. **Secondary methods** are ways you can collect data back in the classroom to help you answer the question, such as research using websites, books or other publications.

You will need to collect both quantitative and qualitative data from these sources. Quantitative data tends to be recorded in numbers, while qualitative data tends to be more observational: descriptions of a place or field sketches. Ensure you record the methods you use in your fieldwork investigation and make them a clear section in your report.

Forming conclusions

Your report should have a clear conclusion at the end of your result and data analysis section. Your conclusion should bring together all the key ideas that you have commented on throughout the report and summarise important findings to your reader. There is no need to repeat quantitative data in this section or to cite references.

Source 5

Coral reefs are particularly susceptible to climate change. We know this because of fieldwork.

Evaluating your fieldwork process

The evaluation section of your report requires critical reflection on your use of data collection methods – sometimes the methods you select are not successful, and need to be improved to help you collect more valuable data. Your evaluation should focus on the following questions.

- What was the most successful **primary method** used, and why? What was the least successful, and why?
- What was the most successful secondary method used, and why? What was the least successful, and why?
- If you were to repeat this study, what would you do differently next time?

Note: Do not comment on your time management, how much fun you had or what you learned in general. This should be a formal report that other professionals could read and learn from.

G5.2

fieldwork task 1

How can we virtually monitor environmental change?

Source 1

Satellites can be useful to track environmental change over time. This image shows smoke from the 2020 California wildfires.

Background to fieldwork approach

Environmental change occurs over time on a variety of scales. It can occur naturally or as a result of human activities. As the global climate warms, the environment is becoming more vulnerable to extreme weather events such as floods, droughts and fires. These hazards may have environmental consequences such as a loss of habitat, tree cover and soil fertility, and an increase in air pollution. Increased natural disasters also affect humans through reduced crop yields, loss of work, damaged infrastructure and direct injury or loss of life.

As technology improves, we are able to use programs, mapping systems and satellites to monitor environmental change and the impacts of those changes. Studying these systems can help us identify areas of improvement. This topic is a good opportunity for you to complete virtual fieldwork and investigate how we can monitor environmental change.

Research question and introduction

The research question you create needs to be focused on monitoring and assessing the impacts of environmental change. As a class, brainstorm these terms and craft a question that you will be able to answer via virtual fieldwork.

Some questions you may consider:

- What is the most successful technological system for monitoring change?
- What monitoring systems need to be improved to ensure human activity does not have vast negative environmental impacts?

You may also choose to select a particular type of environmental change or hazard that interests you. For example:

- How could the 2019–2020 fires have been better monitored to reduce the risk of severe environmental change?
- How is the council monitoring local environmental change as a result of infrastructure upgrades?

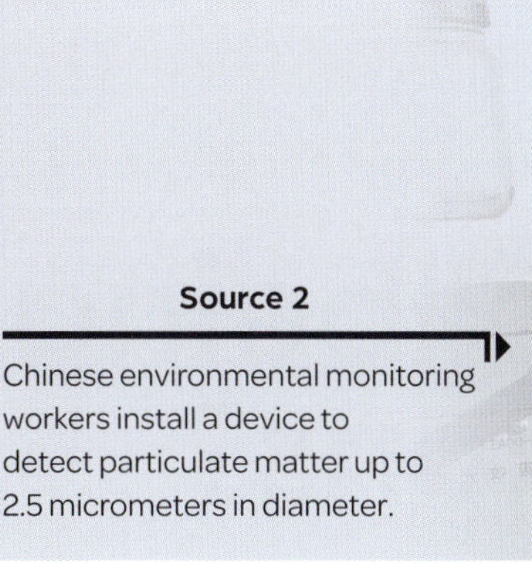

Source 2

Chinese environmental monitoring workers install a device to detect particulate matter up to 2.5 micrometers in diameter.

Once you have a question, you need to create a hypothesis. You may all have slightly different hypotheses, but that will create a more interesting analysis. Complete background research on your chosen question and complete your methodology section. This may be completed before conducting the virtual fieldwork.

Collecting data

As this is a virtual field trip, you need to decide as a class which area of the world you will focus on. You might instead choose to break into research groups and look at different parts of the world.

These weblinks may help when exploring your chosen question.

- Visit Google via http://mea.digital/GHV10_G5_1
- Read an article about time lapse data at http://mea.digital/GHV10_G5_2
- Go on a virtual nature tour from Google Earth at http://mea.digital/GHV10_G5_3
- Discover world human population change over time at the World Population History website at http://mea.digital/GHV10_G5_4
- Research ESRI and GIS: http://mea.digital/GHV10_G5_5
- Finally, you can use NASA and public data systems: http://mea.digital/GHV10_G5_6

Writing a fieldwork report

Using **subheadings** is important in formal reports because it shows the depth of your research and understanding and gives your reader a clear structure to follow. Present the data you collected using tables, graphs, annotated sketches and photographs. Consider how this information helps you to answer the research question.

Your subheadings might be:

- Introduction (background research, question and hypothesis)
- Primary methods
- Secondary methods
- Presentation of data
- Data analysis
- Conclusions
- Evaluation
- References (or bibliography).

What do I do if my hypothesis is 'wrong'?

Once you have completed your research, you need to consider 'does the data support my hypothesis?' Have you begun to change your mind? Your hypothesis was an 'educated guess' you made at the beginning of the fieldwork process. If your hypothesis was supported with your data, great. You can conclude that the research, and the quantitative and qualitative evidence gathered has proven your ideas right! However, if you find that your hypothesis was not supported, that is also useful. Some of the most interesting analyses and discussions come from these situations.

If your hypothesis is 'wrong', it does not mean that your report is 'wrong'. It means that we have more to learn about this topic. This will give you an opportunity to clearly show what you understand about fieldwork, data collection and analysis.

fieldwork task 2

How does human wellbeing vary between regions?

Human wellbeing is affected by a range of factors such as access to resources, income and safety. As a result, patterns of human wellbeing vary between scales and can change over time. Objective data can help us quantify human wellbeing and compare regions according to their health, wealth and level of education. Subjective data can qualify people's happiness and life satisfaction.

You may notice that different suburbs or places in your local government area (LGA) have varied access to infrastructure and resources. Maybe some places are new and are building large developments or undergoing land use changes, while other older areas do not have access to the same facilities. For this fieldwork investigation, you need to research how human wellbeing varies between local regions using both objective and subjective measurements to form conclusions.

Research question and introduction

The research question you create needs to be focused on variations in human wellbeing. As a class, brainstorm these terms and craft a question that you will be able to answer via practical fieldwork.

Some questions you may consider:

- What factors cause variation in human wellbeing on a local scale?
- How has human wellbeing changed over time in the local community?
- How does human wellbeing differ between *place A* and *place B*?
- Is objective or subjective data more useful in identifying variation in human wellbeing on a local scale?

Once you have a question, you need to create a hypothesis. You may all have a slightly different hypotheses, but that will create a more interesting analysis. Complete background research on your chosen question and complete your methodology section. This may be completed prior to conducting the fieldwork.

Source 1

Happiness is dependent on access to jobs and education, as well as recreation and a sustainable environment. The country of Bhutan measures happiness alongside economic measurements.

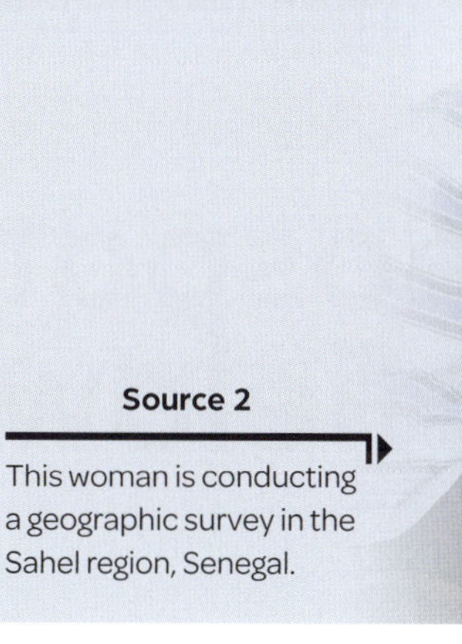

Source 2

This woman is conducting a geographic survey in the Sahel region, Senegal.

Collecting data

The best way to collect subjective data is through surveys. As a class, design a survey that asks locals about their sense of place the services they feel connected to, and provide scales for them to rank their happiness or life satisfaction. Questions should be multiple choice to make data analysis easier. Remember not to ask questions that may make volunteer participants feel uncomfortable, such as asking for income details or whether people are employed. This data can be collected through secondary sources back in the classroom.

Some questions you may present are:

- How would you rate your overall happiness as a resident of *Place A*? Rate answers on a simple scale, such as Source 3.

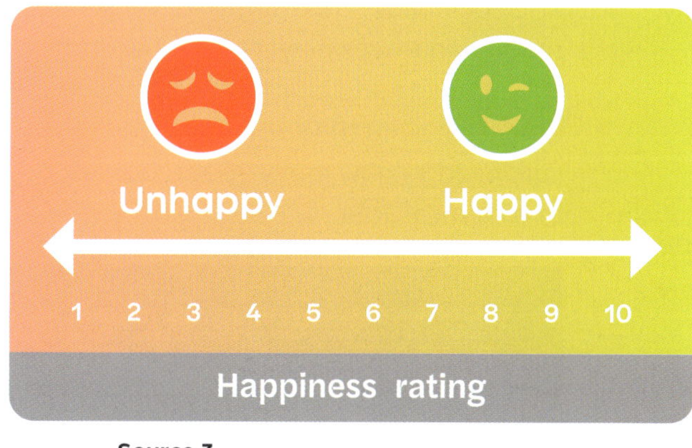

Source 3

A simple happiness rating scale

- Which facilities do you think need improving in your community?
  - **a** Healthcare (hospitals, doctors, family planning services)
  - **b** Access to schooling or education (schools, TAFEs, universities, language centres)
  - **c** Roads and transport services (connectivity)
  - **d** Employment services (job access, training and assistance)

Writing a fieldwork report

Using subheadings, referencing and acknowledging sources is important in formal reports because it shows the depth of your research and understanding and gives your reader a clear structure to follow. Present the data you collected using tables, graphs, annotated sketches and photographs.

Your subheadings might be:

- Introduction (background research, question and hypothesis)
- Primary methods
- Secondary methods
- Presentation of data
- Data analysis
- Conclusions
- Evaluation
- References (or bibliography).

How do I present survey data?

Surveys are most useful when you have a large sample size. Often market research companies will try to survey hundreds or even thousands of people on a particular subject to gain an insight into community needs. For your class, if you all distribute a few surveys each and then share the data, your sample size will be larger and your results will be more reliable.

Once back in the classroom, a few volunteer students may choose to collate the survey data. If you use an online survey such as Google Forms or Survey Monkey, data may automatically collate for you to use at the conclusion of your fieldwork. Survey results are best presented as graphs; however, tables are sometimes more useful if multiple data points are being displayed. Remember to directly refer to your data throughout your report to prove or disprove your hypothesis.

How do I use geospatial technology in my fieldwork?

Using geospatial technology involves engaging with digital platforms to collect, present and analyse data collected in the field. The skills involved in using geospatial technology are not just for fieldwork reporting, but are also useful for work in engineering, surveying, construction, town planning, architecture, mining and entrepreneurship. Use of geospatial technology can be as simple as using your smartphone as part of your fieldwork or accessing free resources online.

Maps and apps

In the past, you may have used fieldwork booklets created by your teacher to structure your research and field experience. While data collection and mapping on paper is helpful, using applications such as Google Maps to track routes and find new locations can be more efficient and effective. Interactive maps allow you to find the latitude and longitude of locations, describe geographical characteristics and define temporal and spatial scales. Geographic apps can also be useful, especially when collating photos or surveys and presenting data in a digital fieldwork portfolio.

- Google Earth
- Google My Maps
- Google Tour Builder
- Google Street View
- Scribble Maps
- National Geographic MapMaker Interactive

- ARCGIS Survey123
- Field counter apps (traffic surveys or other counts)
- Geocam or GPS map camera apps
- My Tracks
- MapMyWalk

Source 1

You may wish to explore some of the apps or virtual maps listed to help you with your fieldwork.

Geocaching

In primary school you may have done some orienteering. In orienteering, tasks are designed to help you develop map and compass skills to locate objects or race to a finish line. Improvements in geospatial technology have allowed orienteering to become an activity known as Geocaching. This digital scavenger hunt encourages players to use an app, GPS tracking and clues to locate hidden objects around the world. You may be surprised by how many geocaches are in your local area.

Source 2

Geocaching has a wide following of geography enthusiasts who use clues and maps to locate treasure around the world.

G6

Geo How-To

The key to success in Geography is understanding fundamental skills and being able to apply them in different situations. This chapter will walk you through some of the key skills for Year 10 and provide examples of how to use them.

Mapping with BOLTSS (NA)

Every map requires BOLTSS in order to be understood by the geographic audience, and some people add (NA) to the acronym to remind them to be neat and accurate. When you construct maps, use BOLTSS like a checklist to ensure you complete your mapping tasks correctly.

B Border

A border is important to show the edges of the mapping field. It provides a clear area inside which to construct your map and makes it appear clear and neat to readers.

O Orientation

An orientation, or compass, helps us understand direction when reading the map. Orientations can be drawn as a 4-point, 8-point or 16-point compass.

L Legend

A legend, or key, is vital to understanding the map. Without a legend, colours and symbols would not make sense and we would not be able to interpret patterns or distributions.

T Title

A title gives us an understanding of what the map is showing. If you are drawing a sketch map, you should also provide a date and time. This allows you to monitor change in a location over time.

Global population density, 2020, measured by number of inhabitants per square kilometre

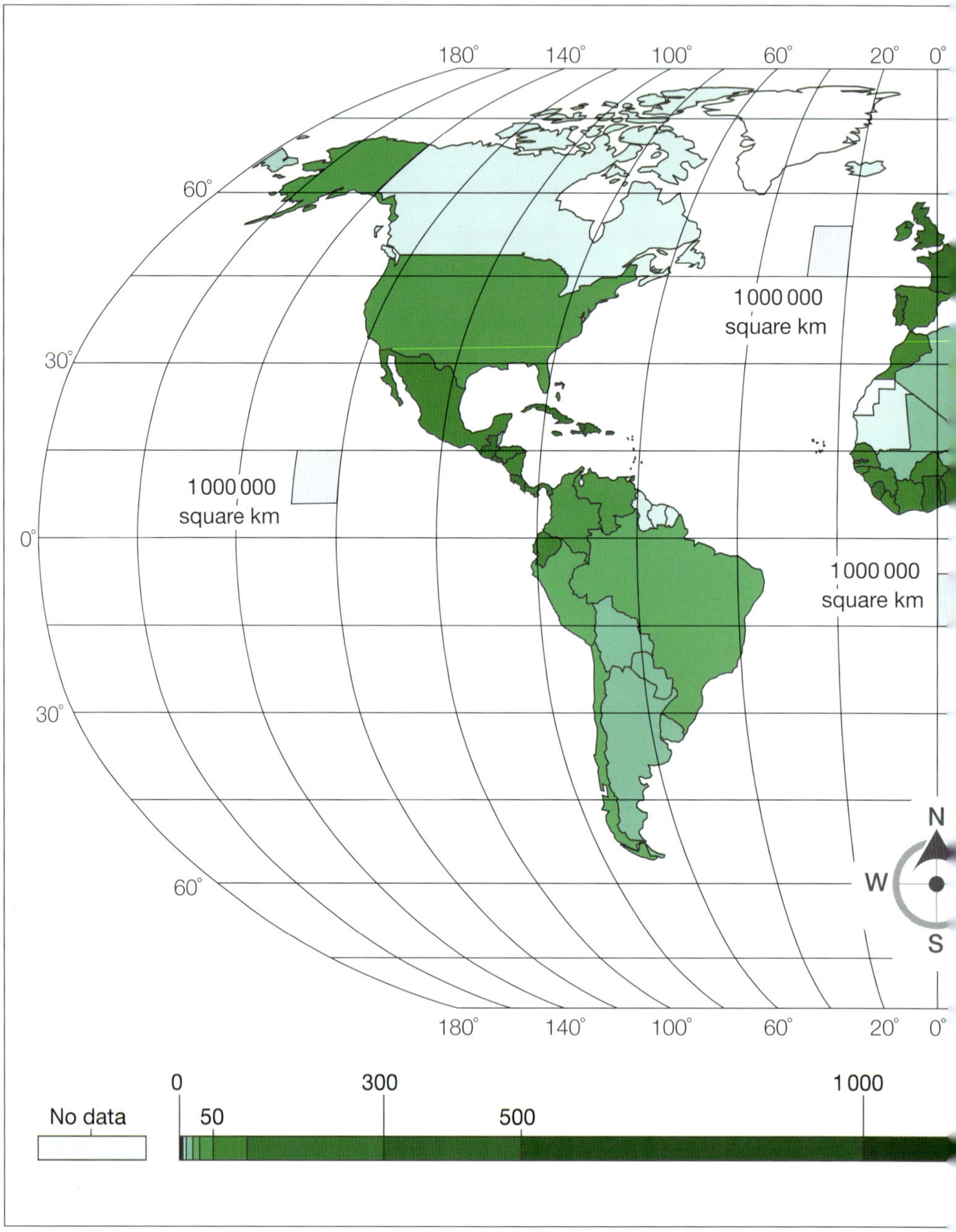

180° 140° 100° 60° 20° 0°

60°

30°

0°

30°

60°

1 000 000 square km

1 000 000 square km

1 000 000 square km

N W S

180° 140° 100° 60° 20° 0°

No data

0 50 300 500 1 000

Source 1

This map features all aspects of BOLTSS.

S Scale

A scale provides us with information on how big something is in real life. While a house on a map may be 2 centimetres across, it may be representing a 15-metre wide three-bedroom home. Scales can come in many forms, such as linear, ratio or fraction.

In the map in Source 1, scale is represented by cubes representing 1 000 000 square kilometres. Different cubes are placed at different positions on the map, depending on the curve of the Earth.

S Source

Always acknowledge the source of the information that you illustrate on your map. The source can also indicate whether the map is reputable or not.

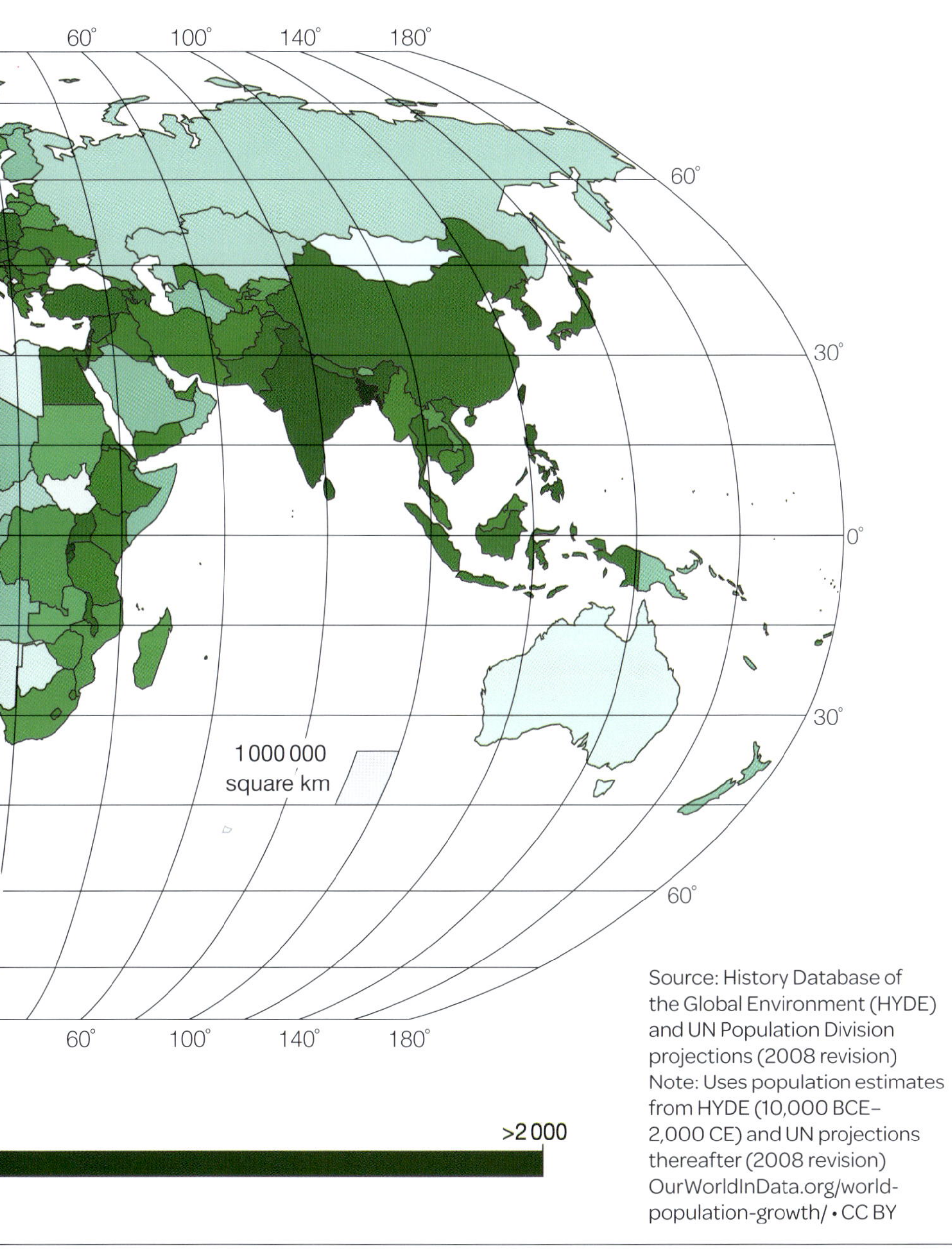

Source: History Database of the Global Environment (HYDE) and UN Population Division projections (2008 revision) Note: Uses population estimates from HYDE (10,000 BCE–2,000 CE) and UN projections thereafter (2008 revision) OurWorldInData.org/world-population-growth/ • CC BY

(NA) Neatness and Accuracy

Some people add the final letters 'NA'. When we read a map, we rely on it being neat and legible and we expect that its data is displayed accurately. Therefore, when you construct a map it is important that you take the time to correctly illustrate the patterns and distributions you see in the data.

Latitude and longitude

The centre of the map features a line of latitude at 0° called the equator. North of the equator is the northern hemisphere, which includes the continents of North America and Europe. South of the equator is the southern hemisphere, where we find Australia. Sometimes, we mistakenly use the word 'above' instead of 'north' or 'below' instead of 'south'. If you say something is 'above' the equator, you are actually saying it is floating in the air over the top of it. If you say something is 'below' the equator, you are describing something buried beneath it. Use directional terms carefully to ensure you are sending people in the right direction.

In Geography we use maps and graphs to understand what is happening around us. In many cases, these resources provide us with information on patterns and spatial distributions of phenomena. When asked to describe a pattern or distribution we follow the PQE formula. The formula PQE helps us describe these patterns and distributions. P stands for pattern, Q stands for quantify and E stands for exception.

P Pattern

A pattern is a trend in the data. When looking for a pattern, read the legend and interpret what the colours or symbols mean. On a graph or map, you may notice that all the data points tend to be clustered in one spot or that there is an uneven distribution of data points. You may need to use the names of places, compass points or lines of latitude to describe where on the map these clusters appear. Descriptive words that may help you describe patterns are: more, less, clustered, even, uneven, highly distributed, north, south, east, west, increase, decrease and fluctuate. When writing a PQE analysis, start the pattern sentence with 'overall'.

Q Quantify

When we quantify our pattern, we need to use numerical data to provide evidence of what we see. You could gain data by using the legend, measuring using the scale or conducting a count. You need to ensure that the data you provide relates directly to the pattern you recorded earlier. When writing a PQE analysis, start the quantification sentence with 'for example'.

E Exception

An exception is a trend on the map or graph that doesn't 'fit in' with our original pattern statement. When you observe an exception, you need to use quantifiable data in order to provide a comparison to our original statement. When writing a PQE analysis, start the exception sentence with 'however'.

Comparative PQEs

As you move towards VCE you may be presented with two maps or graphs and asked to compare patterns and distributions or observe changes over time. We can still use the PQE formula in this case to help us structure a geographical response. There are a number of ways you can form comparisons using PQE, but no matter what approach you choose, you must include examples and quantifiable data in your response.

The most efficient way to form comparisons is to directly compare data within each section of the PQE. In other words, you would use the pattern section of your analysis to highlight key differences in distributions, use data from the map or graph as evidence in your quantification section and then, finally, in the exceptions section, highlight any similarities in the data.

For example, if we were asked: 'Sources 2 and 3: With reference to at least one world region, describe the differences in child mortality rates between 1987 and 2017', you may answer:

Pattern: Overall, the child mortality rate has decreased between 1987 and 2017 in all world regions.

Quantify: Asia has seen one of the largest improvements in child mortality rates. For example, China has seen a reduction from more than 5 per cent to less than 1 per cent child mortality and India has shifted from over 10 per cent to under 5 per cent child mortality.

Exception: MEDCs have consistently had lower child mortality rates over time in comparison to LEDCs. This is particularly true in some African nations where, despite dramatic improvements, child mortality rates are still over 10 per cent in places such as Mali, Chad and Somalia.

PQE example

For example, if we were asked to: 'Refer to Source 3: Describe the distribution of child mortality rates in 2017'. We could answer:

Pattern: Overall in 2017, LEDCs tend to have higher rates of child mortality than MEDCs.

Quantify: For example, Chad (LEDC) has a child mortality rate of over 10 per cent, while Australia (MEDC) has a child mortality rate of less than 1 per cent.

Exception: However, patterns of child mortality vary between LEDCs with places such as Madagascar having a lower child mortality rate (less than 5 per cent) than other nearby countries.

Rates of child mortality, 1987

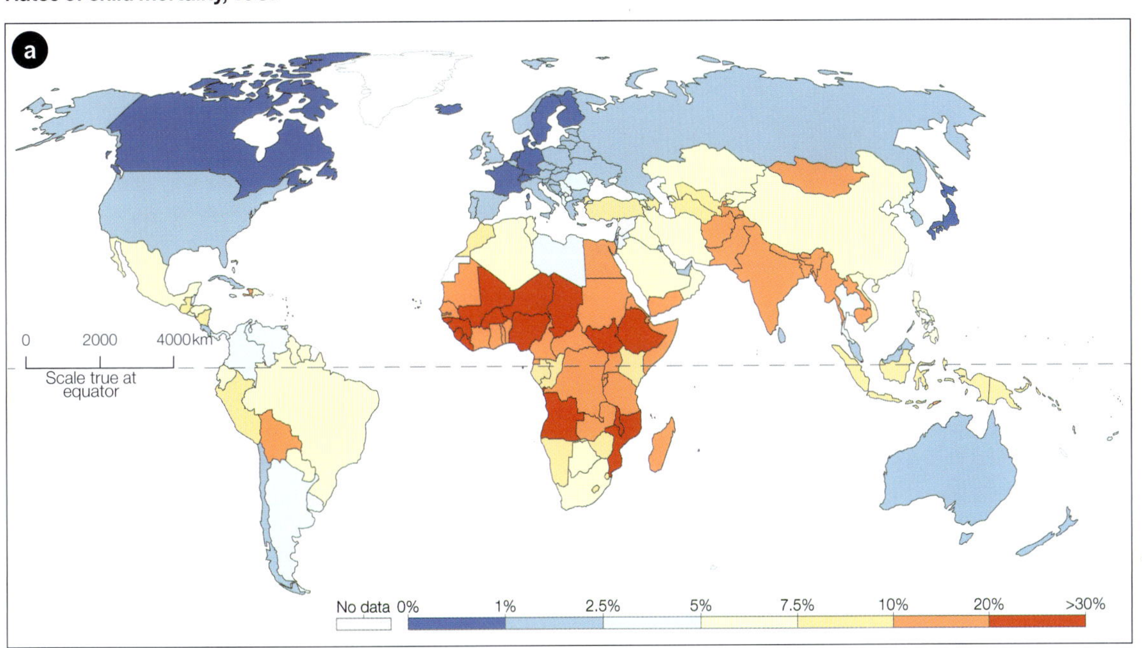

Source: Our World in Data

Source 2

Rates of child mortality in 1987

Rates of child mortality, 2017

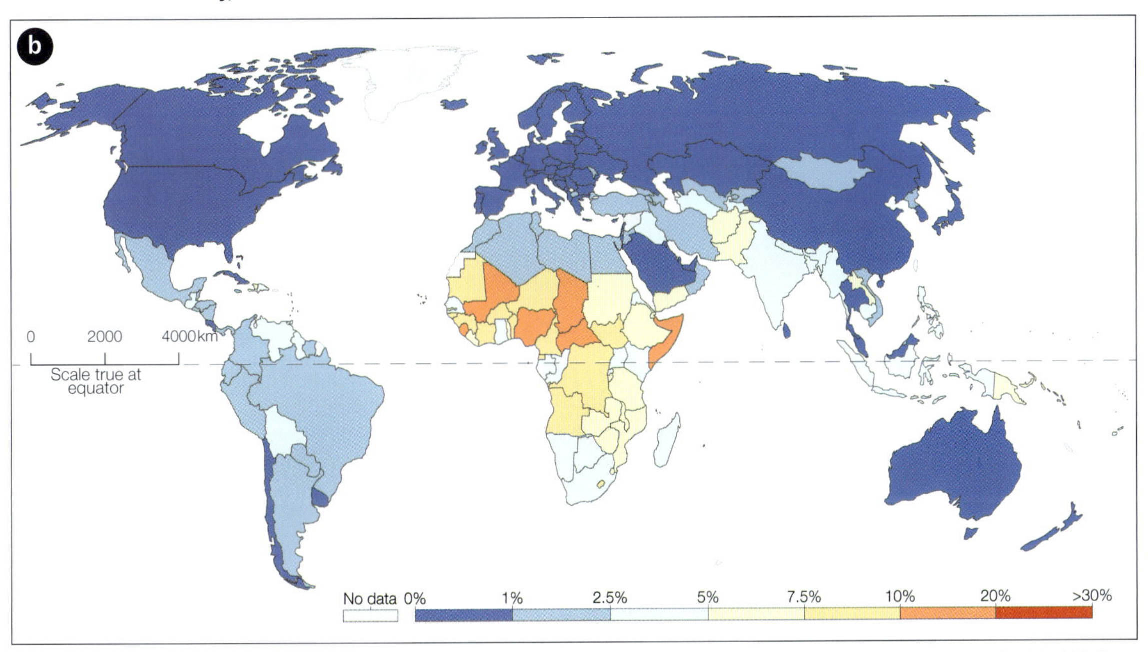

Source: Our World in Data

Source 3

Rates of child mortality in 2017

SHEEPT

SHEEPT is an acronym that helps you remember the reasons why a spatial pattern occurs. It stands for: Social, Historical, Economic, Environmental, Political and Technological.

Source 4

Preparing sago palm logs in Borneo.

S Social

Social factors are anything to do with people. Social factors include population, culture, language and religion.

H Historical

Historical factors are anything to do with our past. Historical events, buildings, people and changes to climate all influence what we see in our world today. Here we see a significant change over time through the loss of native forest for agriculture.

E Economic

Economic factors are those relating to money. In Geography, income, costs of things and how much money is spent can provide us with information on a place.

How do I write a SHEEPT analysis?

SHEEPT is usually used to explain why patterns or distributions may occur in a particular region. It can also be used to expand our thinking when annotating images or considering new geographical content. The text below and on the right is an example of how to write a SHEEPT analysis for an image. The highlighted terms indicate the use of a SHEEPT term. Can you identify all of them? (The analysis does not need to include every term.)

Source 4 is an image of deforestation occurring in a tropical rainforest. In places such as Borneo, deforestation is occurring on a large scale. Trees are predominantly cut down to make way for palm oil plantations. Indonesia and Malaysia produce about 85 per cent of the world's palm oil, which is used in products such as confectionery, sauces and shampoo. In fact, it is present in around 50 per cent of packaged products sold. Deforestation has a range of negative impacts on the environment, from loss of habitat to increased carbon emissions. Deforestation alone is responsible for 10 per cent of the world's carbon footprint. However, many people rely on palm oil for a sustainable income. In Indonesia, traditional crops such as rice earn farmers on average $250 per hectare, whereas mature palm oil plantations can earn up to $2500. As a result, governments are faced with a conflict between managing the rights of traditional land owners, meeting demand for palm oil locally and internationally, and safeguarding the local economy.

E Environmental

Environmental factors are those relating to the natural or human environments on Earth. Humans can manipulate the environment to suit their needs. As our global population grows, the demand for food and resources increases, meaning more land must be cleared for large-scale agriculture such as palm oil plantations.

P Political

Political factors are those to do with the government or leading groups. When we consider political factors, we refer to laws and policies.

T Technological

Technological factors relate to the different kinds of technology that we have access to. This could be in the form of gadgets, geospatial technology, medical technology or even roads, transport or basic machines used in the home.

Sketches and annotating

Field **sketches** are an excellent way of recording data when you are investigating a research question.

Sketches allow you to annotate movement, patterns or any interconnections you see. Field sketching is not a test of your artistic skills – the idea is to record a simplified version of what you can see.

T Title (1)

Create a heading for your sketch that is clear and provides information on the location. You may wish to record both the **absolute** and **relative locations**.

O Orientation (2)

An orientation shows the direction that you were facing when you conducted your sketch. In order to record a correct orientation, you need to use a compass.

A Annotations (3)

Annotations are the most important thing to complete when drawing a field sketch. Annotations allow you to record details about what you see and explain how elements of your drawing relate back to the research question. Ensure that lines pointing to your annotations are completed with a ruler and do not overlap.

DHARAVI, EAST MUMBAI (1)

Bandra District, Maharashtra, India

11 February 2013, 2.35 pm

(3) Haze may indicate high rates of air pollution which will impact people's health.

(5)

(6) Proximity to markets, factories and other potential employment opportunities may have pulled many people to this region over time.

Rubbish and ground cover indicate poor access to sanitation facilities for residents.

TOASTIE will help you remember the key skills when making a field sketch. It stands for Title, Orientation, Annotations, Scale, Time, Information and Edge.

Source 5

An annotated field sketch with accompanying photo

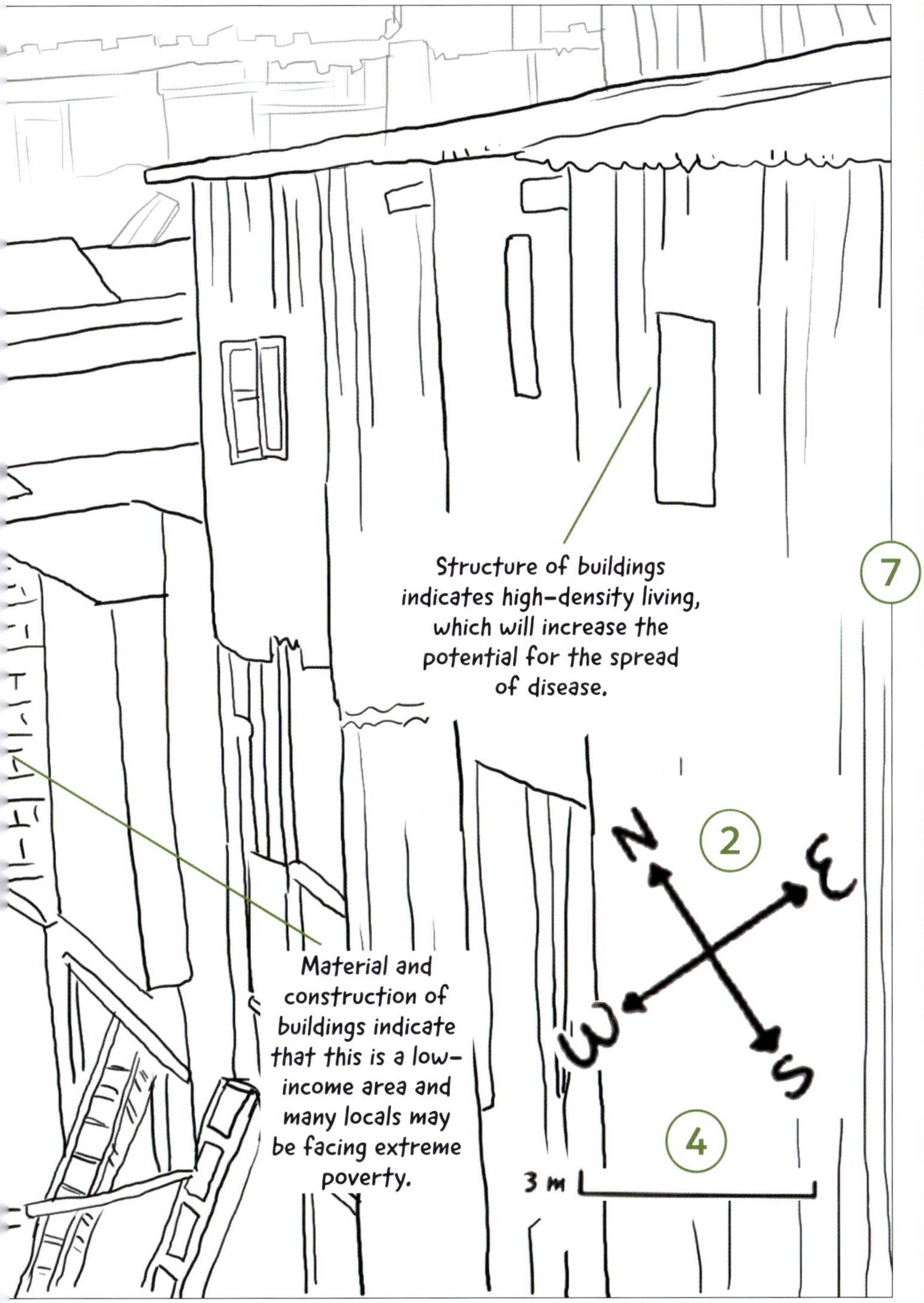

Structure of buildings indicates high-density living, which will increase the potential for the spread of disease.

7

N 2 E W S

Material and construction of buildings indicate that this is a low-income area and many locals may be facing extreme poverty.

4

3 m

S Scale (4)

Most sketches are not to scale. However, all geographical maps and sketches require a scale to give the reader some indication of size. To estimate a scale, use a metre-rule or pace out an area that you have sketched. Then, using a small ruler, identify how large the same area is on your drawing. For example, you may estimate that the path you are looking at is 1 metre wide, and when you measure your drawing of the path it is 1 centimetre wide. Therefore, your rough estimated scale is 1 cm = 1 m.

T Time (5)

By recording the time your sketch was completed, you can analyse how the environment changes over the course of a day, a month or even years!

I Information (6)

Provide more than just one-word annotations on your sketch. Annotations should be at least one sentence and help your reader to identify any patterns.

E Edge 

Draw a border so it is clear where your sketch starts and ends.

Photo essays

A **photo essay** is a way of presenting information visually to show characteristics of a place or process. A photo essay usually includes a series of photos with specific annotations or captions that provide a brief background into the key features of the image or the meaning behind the image choice.

Source 6

A photo essay about global warming

Polar bear waits on sea ice. Sea ice is vital for hunting and refuge from the water.

Sea ice is melting and becoming thinner over time.

Ice has a high albedo rate, which helps to regulate global temperatures.

Researchers track glacial and sea ice change over time.

At any one time, between 1000 and 4000 researchers live and work in Antarctica.

As a result of melting glaciers and warming oceans, we are experiencing global sea level rise.

Sea level rises 5 September, 2017, as major Hurricane Irma approaches the Caribbean island of Sint Maarten

Researchers have attributed more than 50 per cent of the increased sea level to global warming.

Sea level rise will alter coastal processes and increase erosion, putting coastal communities at risk.

In 2014, huge waves demolished sea-wall defences and scattered large stones along the sea-front in Wales, UK.

It is estimated that more than 200 million people will be directly affected by sea level rise on a global scale.

Graphing

Choosing a suitable graph

Graphing is an important way to display geographic information. By using the appropriate graph type we can clearly show patterns and changes over time.

Bar graphs are most suitable for comparing small changes over time that are harder to interpret on a line graph. For example, in Source 7, precipitation is represented in bars as it only ranges from 45–80 mm over the year.

Line graphs are more suitable to show and compare larger changes that occur over longer time periods. For example, in Source 7, temperature is shown to be coolest from June to August, reaching around 10 degrees on average.

Pie charts are best used when you want to show proportions. These charts are not suitable for showing change over time.

When creating a graph, use the acronym **SALTS** to guide you: scale, axis, legend, title and source.

Scale: The scale of your graph will depend on the data you are trying to visualise. To choose your axis scale, identify the lowest and highest value (range), then fill in the numbers in between to mark your data points.

Axis: Each graph has an *x*-axis and *y*-axis. The *x*-axis is the horizontal axis and the *y*-axis is the vertical. Make sure you label each axis!

Legend: A graph often uses colours to represent data. A legend indicates to the audience what these colours mean and how to read the data.

Title: A title lets your audience know what your graph is showing.

Source: When you graph information, you must acknowledge where you obtained your data. By stating the source of your information, the reader knows how reliable it is.

In Source 7, we see a climate graph. Here we are using both bars and lines to display two different patterns that are interconnected over a year. Temperature (degrees Celsius) on the left *y*-axis relates to the red average temperature line; and the amount of precipitation (mm) is shown on the right *y*-axis, which relates to the blue columns. The horizontal axis, or *x*-axis, shows us the months of the year.

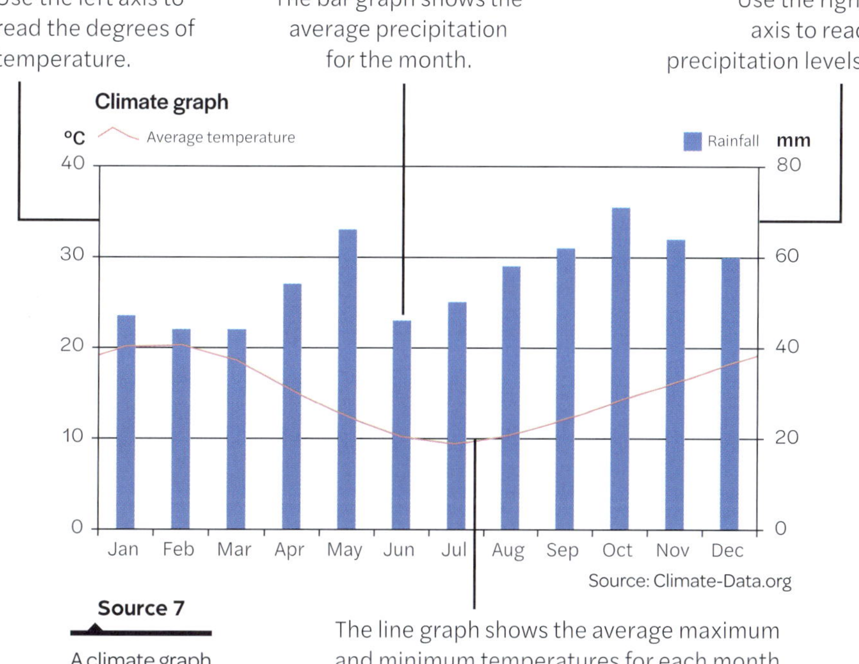

Source 7

A climate graph

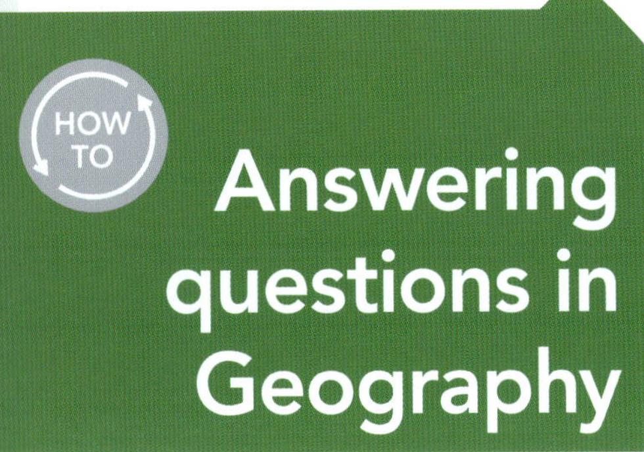

Answering questions in Geography

In Geography there are two main types of questions: those with source data such as maps, graphs and tables, and those without source data. When you are provided with maps, you may be asked to describe, compare or explain patterns that you see. When you are not presented with a source, you may be asked to analyse a statement, reflect on a theoretical concept or apply your understanding through a case study.

As you move towards VCE Geography, geographical questions will move away from the simplistic who, what, where, when, why, how style and you may find that more complex terms are introduced to help you think critically in forming your responses. The following terms appear throughout Good Geography 10. There is no set way to respond to these question terms; however, the following guides may help you formulate a detailed geographic answer.

Question term: *Compare*

Answer approach

Explore the similarities and differences between the data or concepts and then write a short concluding sentence summarising your findings. If comparing data you may use PQE to compare patterns and distributions (see pages 154–5).

Example

Question: *Compare* the trends in weather-related hazard events and geological hazard events from 1980–2000.

From 1980–2000, there has been a steady annual increase in weather-related hazard events, from 45 to 165 annually. Over the same time period, geological hazard events have fluctuated, ranging between 10 and 40 annually. While both show variation, a clear increase in weather-related hazard events is evident.

Question term: *Consider*

Answer approach

Use the data suggested or incorporate the theory or case study suggested into your response.

Example

Question: *Consider* how population growth is likely to affect Melbourne in the future.

Melbourne's current population is around 5 million, but by 2056 is projected to be almost 9 million. This will likely lead to a range of significant changes, including increasing population density, a trend towards medium- and high-density housing, increased traffic congestion and continued urban expansion, particularly in the growth corridors in the southeast, southwest and north.

Source 8

The town of Aasiaat in Greenland during winter.

Question term: *Describe*

Answer approach

When you are provided with a source of data such as a map, graph or table, PQE is the best approach (see pages 154–5).

If you are not provided with data, you need to provide a 'written picture' of what something is or looks like.

Example

Question: *Describe* the geographical characteristics of Greenland.

Greenland is located in the northern hemisphere between the north Atlantic Ocean and Arctic Ocean. It is characterised by large expanses of glacial ice and has a cold climate. It has a small population of just over 56 000 (2018), 88% of whom are Inuit, and it is the least densely populated country in the world.

Question term: *Discuss*

Answer approach

Discuss is a broad question term. You may explore the positives and negatives or different perspectives of a situation or concept. Provide a concluding sentence that summarises your ideas.

Example

Question: *Discuss* the impacts of economic growth on Mongolia.

Over the past two decades, the expansion and increased investment in Mongolia's mineral resources has had a tremendous impact on economic growth, maintaining annual GDP growth above 5 per cent. This has encouraged foreign investment and led to diverse employment opportunities, which has in turn increased living standards.

This rapid growth has been matched by urban expansion, as well as the associated challenges of traffic congestion, rising real estate prices and some of the highest rates of air pollution in the world. While economic growth has brought significant change to Mongolia, it has had both distinctly positive and negative impacts.

Question term: *Explain*

Answer approach

Use elements of SHEEPT to expand your ideas about 'why' something may be happening (see pages 156–7). Use data in your answer as evidence for your explanations.

Example

Question: *Explain* how global warming is contributing to rising sea levels.

Global warming is contributing to rising sea levels because the greenhouse effect prevents heat from dissipating effectively in two direct ways. First, through the melting of land ice, such as glaciers and ice sheets, which increases the volume of water in the Earth's oceans and, second, through thermal ocean expansion, as the increased temperatures cause water particles to expand, thus increasing the relative levels of sea water.

Question term: *List*

Answer approach

Create a series of dot points that highlight key concepts or examples.

Example

Question: *List* four examples of hazard events that may affect Bangladesh.

1. Flooding of the Ganges and Brahmaputra Rivers during spring.
2. Monsoonal downpours that affect low-lying areas.
3. Cyclones that form over the Bay of Bengal and threaten the southern coast.
4. Approximately one half of Bangladesh is located in a high-risk earthquake zone.

Question term: *Outline*

Answer approach

Provide a brief overview of a concept using examples or data.

Example

Question: *Outline* how digital maps are useful in monitoring the spread of COVID-19.

Digital maps allow for a range of stakeholders such as medical personnel, decision makers and members of the community to easily access reliable information. Digital maps are easily updated, allowing for people to report real-time data about the spread of COVID-19 in particular areas.

Question term: *Predict*

Answer approach

This is a crystal ball moment. Using your knowledge of current theories and historical/current examples, you need to suggest what may happen in the future.

Example

Question: *Predict* how global warming will affect sea levels in the future.

Global average temperature has been increasing by around 1.7°C per cent per century since 1970. This global shift in climate is resulting in increased glacial melt of around 1 m per year (2010–2018). As a result, fresh glacial meltwater is resulting in sea level rise.

NOAA recently reported that sea level rise has more than doubled. Unless significant action is taken to reduce the impacts of global warming, it is predicted that sea levels will continue to rise by around 2.5 m by 2100, putting many of our world's largest and most populated cities, including around 85 per cent of Australians (who live within 50 km of the coast), at risk of inundation.

Question term: *Rank*

Answer approach

Put the concepts or examples in order. They may be in order of size or importance.

Example

Question: *Rank* the following countries according to GDP growth: China, Japan, Italy and the USA.

1. China: 6.90%
2. United States: 2.27%
3. Japan: 1.71%
4. Italy: 1.50%

Question term: *Suggest*

Answer approach

Provide a brief proposal or consideration based on a key concept or example.

Example

Question: *Suggest* how human wellbeing could be improved on a local scale.

Human wellbeing could be improved on a local scale through a focus on 'the common good' – ensuring all members of communities are provided with equal access to essential resources such as safe housing, food, medical assistance and open space. Charities and NGOs such as Thread Together in Victoria aim to provide individuals with up-to-date clothing to build confidence and relationships and enhance employment opportunities.

Question term: *Summarise*

Answer approach

Provide a brief overview of a concept or case study using key examples and statistics to provide a clear story.

Example

Question: *Summarise* how human activities are interconnected with environmental change.

As the human population continues to grow, from a total population of around 2 billion in 1920, to around 7.6 billion in 2020, our rapid urbanisation, expansion of agriculture and reliance on industry has meant that we are responsible for significant terrestrial, marine and atmospheric changes. Environmental change brought about by humans generally occurs over a short timescale and is more extensive and less predictable than natural changes.

One of the most significant environmental changes strongly interconnected with human activity is global warming and the enhanced greenhouse effect. Recent studies have revealed that our continued use of fossil fuels and other pollutants has resulted in the highest levels of carbon dioxide in the atmosphere in the past 400 000 years, leading to an average global temperature increase of 0.8°C since 1880. This increased global temperature is responsible for melting glaciers and sea ice, increasing the occurrence of extreme weather events, reducing predictability of growing seasons for crops and loss of biodiversity on a global scale.

Question term: *Evaluate the effectiveness of ...*

Answer approach

When you are looking at a solution to a problem or perhaps a strategy or policy, we can use SWOT to evaluate the effectiveness of that strategy. SHEEPT (pages 156–7) can also help you to expand your ideas here.

Strengths: What are the positive outcomes of the strategy? What does it do well? What are the achievable aims?

Weaknesses: Is there something that the policy does not cover? Are there any clear gaps that need to be addressed? How long will it take to implement?

Opportunities: What improvements could be made? How could it be more effective? What other resources are needed? How sustainable is this policy (economic, environmental and community sustainability)?

Threats: What hurdles need to be overcome (SHEEPT)? What issues might this solution face? How will success be measured? Is it achievable?

Example

Question: *Evaluate the effectiveness of* the Fijian government's relocation strategy as a way of protecting local villagers from the impacts of global warming.

Depending on our management of emissions and future actions, researchers estimate that by 2050, up to 1 billion people worldwide could be displaced as a result of global warming. Fiji is characterised by its 1129 km of coastline, coral reefs and small traditional villages. Fiji is particularly susceptible to sea level rise due to its connection to coastal regions for traditions, housing and income. NOAA recently reported that global sea level has risen by 16–21 cm between 1900 and 2016. As a result, in February 2014, the village of Vunidogoloa was instructed to relocate 2 km inland. Vunidogoloa was the first village to move and, since, another three have been relocated as a result of inundation risk.

While relocation meant that villagers were out of the flood zone and gained access to solar power, rainwater tanks and better facilities, houses were reported to have been built quickly and, therefore, to leak in high-rainfall months. Studies have found that the move has also had a significant impact on the community's sense of place. People have been disconnected from their traditional and sacred lands and access to fishing sites, one of the dominant local food sources.

Further, it is estimated that Fiji will lose up to 5 per cent of its GDP each year as a result of the changing environment, with extreme weather events such as Tropical Cyclone Winston in 2016 only making things worse. After the relocation of Vunidogoloa, the Fijian government established relocation guidelines and identified over 830 communities that will be vulnerable to the impacts of global warming, with 45 of these being earmarked for future relocation.

While Fiji emits less than 1 per cent of the world's carbon output, research has estimated that 4.5 per cent of Fiji's buildings will be flooded by a further 22 cm sea level rise. Therefore, while a strategy of local relocation may be beneficial for some villages in the short term, this may not be an ongoing effective or sustainable solution and, ultimately, if sea levels continue to rise, large Fijian populations may become climate refugees.

Using spatial association in responses

Up until now, you have largely considered how phenomena or patterns are interconnected in time and space. You may have described these interconnections as strong or weak, based on evidence. As you move towards VCE and the use of Key Geographical Concepts, we introduce the new concept of 'spatial association'. Spatial association describes how two phenomena occur in the same place, at the same time for a particular reason.

Spatial associations are also usually described using terms such as 'strong' or 'weak' depending on the degree to which two distributions are related. For example, there is a strong spatial association between annual rainfall and the density of forests, but a weak spatial association between population density and GDP per capita. If two phenomena are not related at all, no spatial association may exist between them.

When answering a question that uses the concept of spatial association, you need to explicitly state the degree of association between the two distributions, before continuing to describe the pattern (using PQE) or explain the connection (using SHEEPT). Remember to provide examples or quantitative evidence to support your ideas.

If asked to describe the spatial association between healthcare expenditure (shown in Source 9) and child mortality (shown in Source 3 on page 155), you might:

1. describe the degree of spatial association
2. complete a PQE (as you are describing distributions on a map, see pages 154–5)
3. ensure you have used evidence and examples to support your spatial association statement.

Healthcare expenditure, 2014

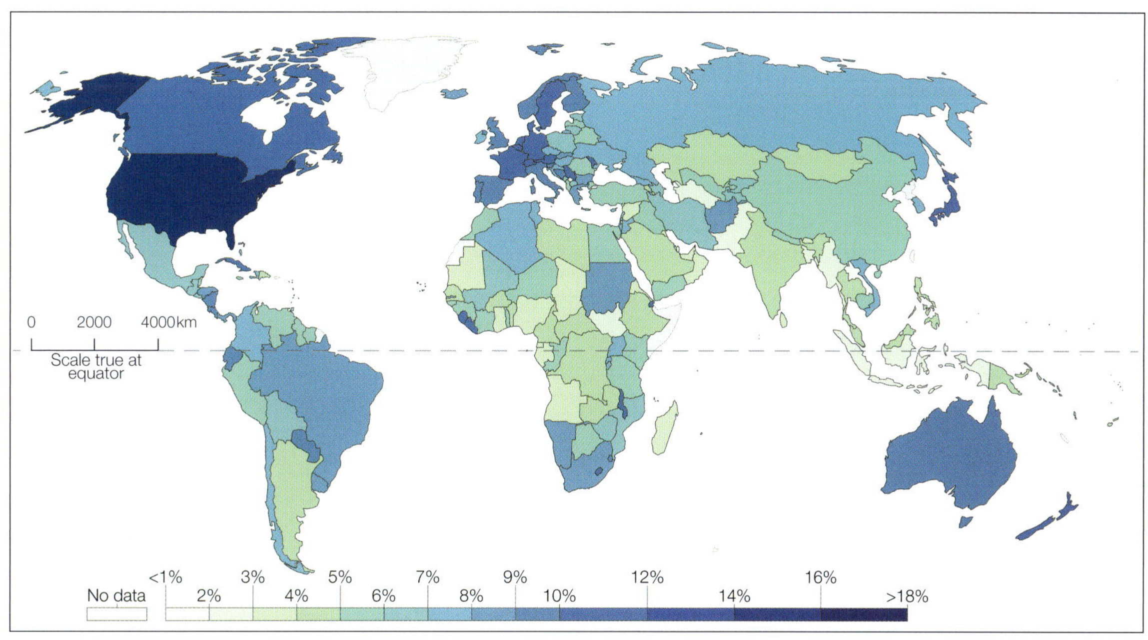

Source: Our World in Data

Source 9

Healthcare expenditure as a share of GDP by country

So, your answer could be:

There is a strong spatial association between healthcare expenditure and child mortality rates on a global scale. Generally, the more a country spends on healthcare, the lower the child mortality rate is in that place.

MEDCs generally have a lower child mortality rate than LEDCs. For example, in 2014, Australia spent over 9 per cent of its GDP on healthcare and, as a result, had a child mortality rate less than 1 per cent. In contrast, Chad spent close to 3 per cent of its GDP on healthcare and, as a result, saw a 13 per cent child mortality rate. However, even though, according to the World Bank, Chad spent more on healthcare (3.62 per cent) than Madagascar (3.04 per cent), Chad still had a higher child mortality rate of 13.5 per cent (in 2014) compared to 5.1 per cent.

If you were asked to explain the strong spatial association between healthcare expenditure and child mortality rates, you might:

1 confirm the degree of spatial association
2 use evidence and examples to support your spatial association statement
3 use some of the SHEEPT factors to explore potential reasons for the relationship.

In this case, you could write:

There is a strong spatial association between healthcare expenditure and child mortality rates on a global scale. Overall, the more a country spent on healthcare, the lower the child mortality rate in that place. For example, in 2014, Australia spent over 9 per cent of its GDP on healthcare and, as a result, had a child mortality rate less than 1 per cent. In contrast, Chad spent close to 3 per cent of its GDP on healthcare and, as a result, saw a 13 per cent child mortality rate.

More expenditure on healthcare in places such as Australia means that more children are able to access medical care when needed. Funding may readily be available to build, update and modernise care facilities and ensure new technologies are being trialled. Free public healthcare is also important for low-income families who would otherwise not be able to access medical advice. Funding for medical research may also mean that countries can focus on prevention of disease and illness through child immunisation programs and parent education.

Glossary

absolute location a precise description of a place's location; for example, an address or geographical coordinates

albedo the reflection of light from a surface; generally speaking, light is reflected away from white surfaces, such as ice sheets, and absorbed by darker surfaces, such as bitumen, causing them to heat

anthropocentric a world view in which humans, as well as their wants and needs, are given priority over the natural world

apartheid political legislation that implemented policies to separate and surpress people based on race

atmospheric referring to the layer of gases that surrounds the planet and is interconnected to processes such as the greenhouse effect.

biodiversity the measure of variation in a community of animals or plants

BOLTSS(NA) the acronym for Border, Orientation, Legend, Title, Scale and Source (Neatness and Accuracy); used when constructing maps

climate average temperature and precipitation in a certain place, over a year

climate change the natural fluctuation in temperature and precipitation over a long time period

cultural burning First Nations techniques of land management using small-scale, controlled fires in specific patches of land to manage human risk and natural fuel build-up

deforestation clearing of a forest to make room for a different land use such as farming

displaced when people must leave their homes on a temporary or permanent basis, often as a result of natural disasters or conflict

displacement when a species or feature has been removed from, or cannot remain, in its natural environment or habitat

ecocentric a world view in which the environment and its species and processes are given priority over the human wants, one in which sustainability is key and environmental footprints are minimal

ember glowing hot coals that can precede or remain after fire

emissions the production or release of gas or radiation during specific natural and human processes, such as manufacturing

endemic species of flora and fauna that are native and restricted to a specific location or region, such as the Tasmanian Devil, which is endemic to Tasmania

enhanced greenhouse effect the disbalance in greenhouse gases in the atmosphere that prevents heat from escaping and causes surface temperature increases

environmental services the benefits provided by nature, such as the sequestering of carbon by trees, or human resources and companies

environmentalists people who focus on the conservation, preservation and promotion of the Earth and its species and resources

exponential refers to an increase that becomes more and more rapid; for example, 1, 2, 4, 8, 16, 32 ...

extinctions the end of a specific species, when no more remain or when the ability to reproduce is no longer possible; prominent examples include the Dodo of Mauritius and the New Zealand Moa.

germination the process by which a seed begins to grow; different species of flora germinate in diverse ways, including through bushfires and flooding

gigafire a fire that burns over an area of more than 1 million acres (just under 404 700 hectares). A fire that burns over 1000 acres (roughly 405 hectares) is called a megafire.

global warming the unnatural rise in the Earth's temperature linked to the increase of fossil fuel use since the Industrial Revolution

green design also known as sustainable design, this seeks to improve the sustainability of residential, commercial and industrial buildings through innovative methods and reducing the consumption of non-natural or non-biodegradable materials

greenhouse gas gases that can absorb or emit solar radiation within the Earth's atmosphere, such as carbon dioxide, methane or nitrous oxide

Gross Domestic Product (GDP) the total value of all goods and services produced over a given period of time, generally a year

hyperinflation refers to runaway inflation, where prices increase rapidly over a short time-scale and the purchasing power of money plummets; often caused by governments printing too much money, which reduces its value

incentive a reward as a motivation to achieve, compete or succeed

innovative agriculture new farming methods that improve yield, reduce the need for water or pesticides and generally support more sustainable practices

Key Geographic Concepts (KGCs) Twelve terms to help think critically about geography: Place, Scale, Distance, Distribution, Movement, Region, Change, Process, Spatial association, Sustainability

less economically developed country (LEDC) a low-income country experiencing severe barriers to development; also referred to as a developing country

malnourishment a lack of proper or adequate nutrition over a period of time leading to negative health impacts

marine related to water and underwater areas, as opposed to terrestrial, such as marine biomes like coral reefs

more economically developed country (MEDC) a country with a strong economy, in which most people have access to good education, health care and employment opportunities

mosaic in the context of cultural burning, refers to burning small scale, low intensity fires in patches within a landscape, rather than encouraging large-scale walls of fire as seen in modern-day back burning

objective factors quantifiable data or information from reliable sources that can be verified

overlay map when layers containing different sets of information or data are created on top of a map

per capita from the Latin 'by head', meaning by person, such as in Gross Domestic Product per capita

photo essay a series of photographs used to present information visually to show characteristics of a place or process

poverty the state of not having enough income to meet one's immediate needs; the UN extends this definition to say 'its manifestations include hunger and malnutrition, limited access to education and other basic services, social discrimination and exclusion, as well as the lack of participation in decision-making'

primary method a data-gathering activity undertaken in the field, such as field sketches

qualitative data non-numerical data based on qualities or characteristics

quantitative data numerical data based on measurements or counts

regeneration the natural process of restoring organic matter to its original state, such as a tree regenerating after a bushfire

relative location a description of where a place is located in the world; for example, describing where your house is in relation to local landmarks

renewable energy an energy source that that can regenerate or replenish within a human timescale

research question an idea to be investigated or a problem to be solved through research

resource curse the phenomena where countries blessed with natural resources are also riddled with political, social or economic issues; also known as the *paradox of plenty*.

sea walls manmade coastal defences designed to mitigate the impact of wave activity on the coastline and on human resources

sense of place the ties or emotional connections that people or communities hold to specific locations; this may originate from personal experience, as well as social, cultural or religious reasons

secondary method a data-gathering activity undertaken outside of field studies, such as research; data collected by others outside of your research group

SHEEPT the acronym for Social, Historical, Environmental, Economic, Political and Technological factors; used when explaining the reasons why a spatial pattern occurs

sketch a simple drawing made to record data when in the field

spatial technology a computer system that interacts with real-world locations in some way

subjective measures data, feedback or qualification based on opinion and experience; open to wider interpretation, bias and variation

subheadings smaller headings to distinguish separate sections or topics

sustainability ecological balance in terms of replenishment and consumption

tailings the waste remaining after separating the valuable minerals from the uneconomic minerals of an ore during mining

terrestrial related to land or earth, as opposed to marine, such as terrestrial biomes like forests or deserts

TOASTIE the acronym for Title, Orientation, Annotations, Scale, Time, Information and Edge; used when making a field sketch

trajectory the curved path an object or line follows once released

urbanisation the process of rural land areas becoming increasingly urban

Index

D

E

F

G

R

S

Acknowledgements

The author and publisher are grateful to the following for permission to reproduce copyright material:

PHOTOGRAPHS: AAP/Reuters Graphic, **71** (bottom); Alamy Stock Photo/ Aaron Amat, **97**, /Abaca Press, **81**, /ajith achuthan, **40–41** (bottom), /Album, **67** (bottom), /Aleksandr Bryliaev, **150** (left), /Anas Alkharboutli/dpa, **123** (bottom), /Anna Berkut , Contents (bottom left), **141**, /Ashley Cooper pics, **39** (top), /Avalon.red , **23** (centre), /Ben Molyneux, **72–73** (bottom), /Bjanka Kadic, **27** (top), /Bleyer/FRIEDRICHSMEIER ARCHIVE, **132** (right), /blickwinkel, **170–75**, /Canvan Images, **43–44**, /christopher jones, **150** (right), /Christopher Scott/Alamy Live News, **135**, /Chun Ju Wu, Contents (top left), **1**, /CSI Productions, **91**, /david hancock, Contents (background), /David Hewison, **102** (bottom), /Dennis Schmelz/mauritius images GmbH, **130–31**, /Dima, **27** (bottom), /DisobeyArt, **94–95**, /dpa picture alliance, **60–61**, /Ebrahem Khadir/UPI, **124–25**, /Edward Tovmassian, **59**, /Ezra Acayan/NurPhoto/ZUMAPRESS.com/Alamy Live News, **128**, /Foto Arena LTDA/Alamy Live News, **64–65** (bottom), /frans lemmens, **50–51** (bottom), /Gelia , **16–17** (bottom), /Geopix, **146**, / Graham M. Lawrence/Alamy Live News, **160** (bottom right), /guy oliver, **92**, /Hans Alm/Folio Images, **116–17**, /Howard Davies, **107** (top), /Iain Masterton/Alamy Live News, **2**, /imagegallery**2**/Alamy Live News, **44**, /Imaginechina Limited, **147**, /Ismael Mohamad/UPI, **79**, /Jacek Lasa, **150** (centre), /Jacek Sopotnicki , **134** (top), /James Schaedig, **36–37** (bottom), /Jamie Pham, **40** (top), /JHeinimann, **133**, /Jivko Konstantinov, **160** (top right), /John R. Kreul/Independent Picture Service, **148**, /JONATHAN AYRES, Contents (top right), **47**, /Justin Hannaford, **49**, /Ken Duffney, **37** (top), /Kristian Buus, **51** (inset), /Leigh Henningham, Contents (centre left), **77**, /M G Photography, **160** (bottom left), /Marc Anderson, **26**, /Marco Palladino, **158**, /Marco Taliani de Marchio, **101**, /Marie Bienaimé/BSIP, **121**, /Marko Prešlenkov, **168–69**, /MediaPunch Inc, **108**, /Miniloc/mauritius images GmbH, **126–27**, /Olaf Krüger/ ImageBROKER, **160** (top left), /OMONIYI AYEDUN OLUBUNMI, **52–53**, /P&F Photography, **35**, /PEROUSSE Bruno/hemis.fr, **93**, /Peter Hermes Furian, **51** (top), /Peter Mundy, **83**, /Philip Game, **102** (top), **104**, /Photo Researchers/ Science History Images , **86** (bottom), /Piter Lenk, **69**, /Prisma Archivo, **123** (top), /Putu Artana, **4–5**, /Rafael Ben-Ari , **54–55** (bottom), /richard sowersby, **25**, /Ryhor Bruyeu, **115**, /Sarah Jane, **107** (bottom), /SimonStone/ Science Photo Library, **22**, /Seaphotoart, Contents (top centre), **11**, /Sebastian Kahnert/dpa-Zentralbild/ZB/dpa/Alamy Live News, Contents (centre right), **111**, /Simon Stone/Science Photo Library, **14** (bottom right), **15** (bottom), **24**, **156–57**, /SPANI Arnaud/hemis.fr, **149**, /Stephen Dwyer, **87**, /steve sadler images, **20**, /Studio CP/Cultura Creative RF, Contents (bottom right), **151**, /Sue Cunningham Photographic, **65**, /Sunshine Seeds, **132** (left), /THP Creative, **68**, /Travelscape Images, **145**, /Urikiri-Shashin-Kan, **118**, /warmheart-MW, **134** (bottom), /WILDLIFE GmbH, **33**, /WJ, **23** (bottom), /Zoonar/Iryna Volina, **127**, /Zoonar/Sergey Mayorov, **66**; Alamy Stock Vector/Naschy, **18**, /Rainer Lesniewski , **54**, **114**, /YAY Media AS , **122**; Fitted for Work, Photo taken by Fi Mims, **136**; Getty Images/AFP/Stringer, **62–63**, /Lonely Planet, **71** (top); Used with the permission of Inter IKEA Systems B.V., **73** (top); iStock/AlexanderYershov, **10**, **46**, **76**, **110**, **140** (arch icon), /Beverley Van Praagh, **13**, /Blueastro, **80**, /David Lewis, **113** (top right), /fivepointsix, **113** (top left), /kowition, **90**, /VectorMine, **86** (top); Nansen Initiative, **56**; NASA Earth Observatory, **14** (top, bottom left), **15** (top left, top right); Shutterstock/Rainer Lesniewski, **67** (top). Unsplash/Jesse Schoff, **75**, /Jezael Melgoza, **176**, /Johannes Plenio, **45**, /Katt Yukawa, **139**, /Matt Collamer, **105**, /Nina Strehl, **100**, /Victor Garcia, **39** (bottom), /Victória Kubiaki, **109**, /Visit Greenland, **162–63**.

OTHER MATERIAL: Graph, Reprinted with permission from Roland Geyer, Jenna R. Jambeck, Kara Lavender Law, Science Advances 19 Jul 2017: Vol. 3, no. 7, e1700782 DOI: 10.1126/sciadv.1700782 © The Authors, some rights reserved; exclusive licensee AAAS. This work is distributed under CC BY-NC creativecommons.org/licenses/by-nc/4.0/, **129**; Extract from 'I collect and sell anything': Waste-picking a means to an end for Philippines' poor', by Shirley Escalante, 4 Sept 2016. Reproduced by permission of the Australian Broadcasting Corporation – Library Sales © 2016 ABC", **128**; Table, Australian Bureau of Statistics. Licensed under CC BY 4.0 Licence creativecommons.org/licenses/by/4.0/, **105**; ACOSS/UNSW Poverty and Inequality Partnership, povertyandinequality.acoss.org.au, **137** (centre); Artwork, Adnate, Contents (centre left), **77**; Cartoon, Frits Ahlefeldt-Laurvig, **101**; Map, sourced from Cal Fire. Map screenshot images are the intellectual property of Esri and is used herein with permission. Copyright © 2021 Esri and its licensors. All rights reserved, **58**; Map, CDC, **138** (top and centre right); Infographic, Climate Council, **34**; Map © Commonwealth of Australia 2020, Bureau of Meteorology. Licensed under CC BY 3.0 AU Licence creativecommons.org/licenses/by/3.0/au/, **30** (bottom left, bottom right); Map © Commonwealth of Australia (Murray–Darling Basin Authority). Licensed under CC BY 4.0 Licence, **68**; Fitted for Work Banner, Designed by Isabel Owe-Young, photos sourced from Canva, **137** (top); Screenshot, Google Scholar is a trademark of Google LLC and this book is not endorsed by or affiliated with Google in any way, **142**; Map, Reproduced with permission from I. Friis: "The Flora of Woody Plants and Vegetation on the Horn of Africa", carlsbergfondet.dk/en/Forskningsaktiviteter/Research-Projects/Other-Research-Projects/Ib-Friis_The-flora-of-woody-plants-and-vegetation-on-the-Horn-of-Africa, **61**; Map, Hannah Ker, **14-5** (centre); Graph, IEA (2020), World Energy Outlook. All rights reserved, **35**; Used with the permission of Inter IKEA Systems B.V., **72**; Map data source: Lakhani, Ali; Wollersheim, Dennis (2021): Parklands within a 5km Radius of Metropolitan Melbourne Experiencing Stage 4 Restrictions. La Trobe. Dataset. doi.org/10.26181/6041a8ae56c87. Released under CC BY 4.0 Licence, **103**; Map, Levi Westerveld, www.grida.no/resources/13240, **57**; Magazine cover, Copyright Metro Guide Publishing: A Division of Advocate Media. Nova Scotia, Canada, **55** (top); Graph, NASA Earth Observatory, using data from Williams, Park A., et al. (2019), **59**; Map, Runting, R. K. et al, Alternative futures for Borneo show the value of integrating economic and conservation targets across borders. Nat. Commun, 6:6819 doi: 10.1038/ncomms7819 (2015). Licensed under CC BY 4.0 Licence, **74** (top); Adapted from Cheung, F., Kube, A., Tay, L. et al. The impact of the Syrian conflict on population well-being. Nat Commun 11, 3899 (2020). doi.org/10.1038/s41467-020-17369-0, Figures 2, 4. Licensed under CC BY 4.0 Licence, Map, **124** (bottom), Graph, **125**; Map © Nic Marks see happyplanetindex.org, **82–3**; Cartoon, courtesy of njlifehacks.com, **97**; NOAA Climate.gov, Screenshot, **6**, Graph, **20**, **23** (top), **33** (top), Map, **33** (bottom); Infographic, OECD (2020), "How's Life in Japan?", in How's Life? 2020: Measuring Well-being, OECD Publishing, Paris, doi.org/10.1787/800829e4-en, **121**; Map, Hannah Ritchie and Max Roser (2013) – "Land Use". Published online at ourworldindata.org/land-use. Data source: World Bank – World Development Indicators, **17** (top left); Map, Max Roser, Hannah Ritchie and Esteban Ortiz-Ospina (2013) – "World Population Growth". Published online at ourworldindata.org/world-population-growth. Data source: World Bank – World Development Indicators, **17** (top right); Graph, Hannah Ritchie and Max Roser (2020) – "Renewable Energy". Published online at ourworldindata.org/renewable-energy. Data source: BNEP and World Bank, **41**; Graph, Hannah Ritchie and Max Roser (2020) – "CO2 emissions". Published online at ourworldindata.org/co2-emissions Data source: Our World in Data based on Global Carbon Project; BP; Maddison; UNWPP, **44** (top right); Graph, Hannah Ritchie and Max Roser (2020) – "Agricultural Production". Published online at ourworldindata.org/agricultural-production. Data source: Food and Agriculture Organization of the United Nations (FAO) (2020), **74** (bottom left); Graph, Hannah Ritchie and Max Roser (2013) – "Crop Yields". Published online at ourworldindata.org/crop-yields. Data source: UN Food and Agriculture Organization (FAO), **74** (bottom right); Map, Hannah Ritchie and Max Roser (2019) – "Access to Energy". Published online at ourworldindata.org/energy-access. Data source: World Bank – World Development Indicators, **87**; Map, Esteban Ortiz-Ospina and Max Roser (2018) – "Economic inequality by gender". Published online at ourworldindata.org/economic-inequality-by-gender. Data source: UN Development Programme, **89**; Map, Max Roser (2014) – "Human Development Index (HDI)". Published online at ourworldindata.org/

human-development-index. Data source: United Nations Development Programme (UNDP), **84**; Map and graph, Esteban Ortiz-Ospina and Max Roser (2013) – "Happiness and Life Satisfaction". Published online at ourworldindata.org/happiness-and-life-satisfaction. Map Data source: World Happiness Report 2019, **108** (top), Graph data source: World Bank – World Development Indicators, **96**; Graph, Max Roser and Esteban Ortiz-Ospina (2013) – "Global Extreme Poverty". Published online at ourworldindata.org/extreme-poverty. Data source: World Bank – World Development Indicators, **108** (bottom); Map, Max Roser, Hannah Ritchie and Esteban Ortiz-Ospina (2013) – "World Population Growth". Published online at ourworldindata.org/world-population-growth. Data source: History Database of the Global Environment (HYDE) and UN Population Division (2008 Revision), **152–3**; Map, Max Roser, Hannah Ritchie and Bernadeta Dadonaite (2013) – "Child and Infant Mortality". Published online at ourworldindata.org/child-mortality. Data source: UN Inter-agency Group for Child Mortality Estimation, **155** (top, bottom); Map, Esteban Ortiz-Ospina and Max Roser (2017) – "Financing Healthcare". Published online at ourworldindata.org/financing-healthcare. Data source: World Bank – World Development Indicators, **167**. All Our World In Data material licensed under a Licensed under CC BY 4.0 Licence; Map, from Raffael M. Tófoli, Rosa M. Dias, Gustavo H. Zaia Alves, David J. Hoeinghaus, Luiz C. Gomes, Matheus T. Baumgartner, Angelo A. Agostinho, Gold at what cost? Another megaproject threatens biodiversity in the Amazon, Perspectives in Ecology and Conservation, Volume 15, Issue 2, 2017, Pages 129-131, ISSN 2530-0644, doi.org/10.1016/j.pecon.2017.06.003. Reproduced with permission, **64**; Map, Figure 1 (A) The ensemble-mean difference in annual temperature between the CMIP5 Historical and Natural forcing experiments during the IPCC's historical baseline period (1986–2005), from Global warming has increased global economic inequality, Noah S. Diffenbaugh and Marshall Burke, PNAS May 14, 2019 116 (20) 9808-9813; first published April 22, 2019 doi.org/10.1073/pnas.1816020116, **44** (top left); PopulationPyramid.net, made available under a Creative Commons license CC BY 3.0 IGO: creativecommons.org/licenses/by/3.0/igo/, **119** (left, right), **124** (top left, top right, centre); Graph, Greyling, Talita, Stephanie Rossouw, and Tamanna Adhikari. "A Tale of Three Countries: What is the Relationship Between COVID-19, Lockdown and Happiness?" South African Journal of Economics 89, no. 1 (2021): 25-43, **95** (top); Graph, Statista, www.statista.com/chart/17428/happiest-countries-in-the-world/ Licensed under Creative Commons Licence CC BY-ND 3.0, **116**; From The Sustainable Development Goals Report 2020, by Department of Economic and Social Affairs ©2020 United Nations. Reprinted with the permission of the United Nations, www.un.org/sustainabledevelopment/ The content of this publication has not been approved by the United Nations and does not reflect the views of the United Nations or its officials or Member States, **98**, **99**; UNDP Human Development Reports. Licensed under Creative Commons Attribution 3.0 IGO, **85**, **88**, **131**; Map, U.S. Census Bureau, **138** (bottom left); Diagram, USGS, **56**; Logo, courtesy of threadtogether.org, **137** (bottom); Map, adapted from Asus2004, Wikimedia, **133**; The World Bank: Dataset name: What A Waste Global Database. Licensed under CC BY 4.0 Licence, **28**; Diagram, based on World Health Organization. Ecosystem goods and services for health, who.int/globalchange/ecosystems/en/, accessed 12 March 2021, **91**; Table, Data sources: International Comparison Program, World Development Indicators database, World Bank, Eurostat-OECD PPP Programme; World Bank national accounts data, and OECD National Accounts data files. Licensed under CC BY 4.0 Licence, **134**; Map, Worldmapper, **7** (top).

While every care has been taken to trace and acknowledge copyright, the publisher tenders their apologies for any accidental infringement where copyright has proved untraceable. They would be pleased to come to a suitable arrangement with the rightful owner in each case.

good geography